MOLECULAR BIOLOGY

MOLECULAR BIOLOGY

Dr. Aparna Bhattacharya

RAJAT PUBLICATIONS
New Delhi

Molecular Biology
by *Dr. Aparna Bhattacharya*

This edition is published by Rajat Publications
4740/23, Ansari Road Daryaganj,
New Delhi - 110002

ISBN: 978-81-7880-421-7

Edition 2026

Printed and bound in India

RAJAT
PUBLICATIONS

4740/23, Ansari Road, Daryaganj,
New Delhi- 110002 (India)
Mob:- +91-9811484973, +91-9811105389
E-mail: rajat_publication@yahoo.com

Contents

Preface

Molecular Biology has become the vital aspect of studies in modern life sciences. The branch deals with study of vital molecules of life like DNA, RNA, and Protein etc. It is contributing significantly in understanding the scientific basis of life at Molecular level. Molecular Biology have made emergence of new disciplines like Recombinant DNA technology, Bioinformatics, Proteomics, Regenerative Medicines and Molecular farming etc.

It basically explains the relationship between the structure and function of Biological Molecules and how these relationships contribute to the operation and control of Biochemical process. The Biological molecules of principles importance are DNA, RNA and Protein and the process of Replication, Transcription and Translation. The progress of understanding all vital activities in Organisms are completely depends on new experimental techniques available in Molecular biology. The principal applications of Molecular biology are development of new and safe products such as Therapeutics, Vaccines, Foodstuff's, Transgenic organisms, Diagnosis of diseases and Gene therapy.

This book illuminates the contribution of Molecular biology to Biotechnological advancement. The students of Undergraduate and Postgraduate can exploit this volume to success in universities examination and career development.

Chapter 1

Introduction

Biology is the science devoted to the study of living objects. Two Greek words, bios (life) and logos (discourse), explain the term "biology" as meaning study of living creatures. Since life manifests itself in various forms i.e. plant and animal forms, biology is the study of both. The study of plants is termed as botany or plant biology, while the study of animals is called zoology or animal biology.

Biology also comprises of the cellular basis of living things for the activities of life. It consists of energy metabolism and also of genetics which forms the basis of inheritance in organisms. Living things have been studied since primitive times. Primitive man had to find out what could be utilized for food, shelter and clothing, which plants would be useful as medicines and which animals he could use to get work done for him. The science of biology thus arose out of basic human needs. Biology also includes evolutionary relationships between organisms. Biology tells us about the routine activities of microorganisms, plants and animals and also the relation between their structures and functions. Many sub-divisions and special areas of biology exist but biology can conveniently be divided into two categories:

- Practical
- Theoretical.
  - Practical biology consists of plant breeding, medical science, wildlife management, agriculture.
  - Theoretical biology consists of Physiology (the study of the functions of living things),

Microbiology (the study of microscopic organisms), Biochemistry (the study of the chemical makeup of organisms), Ecology (the study of populations and their environment) and Taxonomy (categorization).

MOLECULAR BIOLOGY

Living things are made up of cells which form the basic unit of life. The cell is, in turn, made up of molecules and atoms. The cell usually cannot be seen with the naked eye. Each cell performs certain functions. Some functioning cells come together organize to form a tissue. It is at this stage that the complex structuring of living things begins. A group of tissues, when organized, form organs such as the brain, the heart, etc. When a number of organs work together, they compose an organ system. So an organism is a composite union of various organ systems. The combination of several organisms is called population, which, in turn, forms a community, leading to the creation of a Biosphere. represents the structural complexity of the Biosphere.

Metabolism

An organism is basically made up of six basic elements: C, H, O, N, P and Ca. These make up about 96% of the weight of an organism. To keep life intact there is an exchange of chemical matter from cell to environment and environment to cell. Organic matter from the environment is absorbed by the cell and utilized for preparing energy. The energy thus prepared is used to perform the functions of the cell. Unlike in nonliving things, a number of physical and chemical changes take place in the body of the organism.

The sum total of physical and chemical changes in an organism is called metabolism. Metabolism includes anabolism in which complex organic substances are built up and energy is stored, and catabolism in which the complex organic substances are broken down to release energy. These changes of conservation and release of energy continuously take place in the organism. This energy is basically in the form of chemical energy. Nonliving matter does not show this kind of exchange of energy.

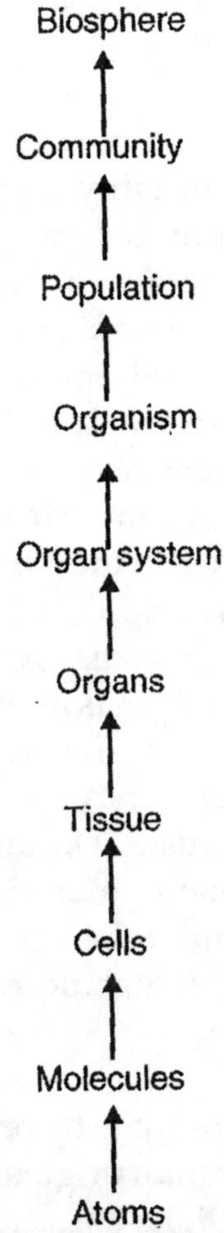

Fig. Structural Complexity of the Biosphere.

Growth

In metabolism we have seen that organic matter is collected from the environment and processed in the cell. In the process of growth, the cell organizes this material into its own structure. During growth the cell spends some amount of the energy created in the process of metabolism. Every organ has its own pattern of growth. This pattern depends upon the tissue of the organ. Growth of different organs leads to physical growth. Food is converted into organic matter which is then utilized for the growth of the body. Nonliving organisms do not show these characteristics.

Reproduction

A living thing has the ability to procreate when it matures through the process of reproduction. There are differences in the reproductive functions of higher animals and lower animals. In lower animals (as in unicellular animals like amebas and bacteria), reproduction merely extends the process of growth. For instance, after maturity, a single bacterium splits into two bacteria, in six stages. This process is called asexual reproduction. In asexual reproduction only one parent is involved. So generally the produced cells are identical to the parent cell.

In asexual reproduction, the rate of production of offspring is faster than in sexual reproduction. In higher or more complex organisms there is another type of reproduction called sexual reproduction. In sexual reproduction, two parents of opposite gender type are responsible for the formation of a new organism. During sexual reproduction, a new combination of genetic material (D.N.A.) occurs. This process is complex and takes more time than asexual reproduction. Nonliving things do not reproduce.

Responsiveness to Stimuli

All living things are able to respond to stimuli in the external environment. Stimuli originate from different sources.

- Physical : light, heat, temperature, sound.
- Chemical : acids and alkali.
- Mechanical : friction, pressure.

The body has specialized tissue or higher centres to detect stimuli, such as eyes for light, ears for auditory stimuli, the nose for olfactory stimuli, the tongue for taste, the skin for touch, etc. An organism must co-ordinate well to give an appropriate response to the stimuli from the environment. All systems of the body help in this co-ordination, like nerves from the nervous system, and certain chemical regulators called hormones from the endocrine system. This simultaneously produces an effect of co-ordination and acts as a sensing system. To respond to the stimuli, the organism has certain efferent units like muscles, glands, hair, etc. The process of response involves the use of energy.

The responses are of two types:

- Generalized,
- Protective.

Protective responses help to promote survival and thus sustain life. A generalized response contribute to the behaviour patterns of an organism. For certain classes, the response occurs with a definite pattern which is innate to the organism. Nonliving things don't respond to stimuli as living things do.

Evolution

Living organisms normally interact with the environment for their daily requirements. There are always changes going on in the environment. To adapt to these changes, there are genetic changes in the animal. The process of adaptation in a new generation animal in the population is called evolution. Evolution allows for changes in interactions between the organism and its environment. The evolved organism is more capable of adapting to changes in the environment. This adaptation in the organism leads to the formation of a new species and also newer organisms within the species. So with evolution, a new species is formed but certain characteristics from the previous generation remain intact.

Ecology

Ecology is the study of the relationships between the environment and the organism, and between various organisms. Due to interaction with the environment, the organism and its environment continuously influence each other. Animate organisms can migrate to another place if the environment is not suited to their survival in a given region. For instance, frogs cannot live in water if there are too many crocodiles on their side of the pond. Nonliving things cannot change their environment.

SCIENTIFIC METHODOLOGY

To study the sciences, including biology, a person must adopt a certain manner of thinking about questions, The methods used to answer questions about biology require a scientist to think about all the possibilities that could explain an observation. The scientist then refers to the research that other scientists and experts have done which might help him or her to better understand the problem. Then experiments are designed which will rule out all the possibilities exept one. If all possibilities exept one are ruled out, then the remaining possibility must be true explanation. Scientific method consists of the following steps.

Observation

The scientist objectively (without forming opinions yet) observes a particular phenomenon in nature or in the labratory. A problem based on that observation is declared. The scientist then formulates the required steps to solve the problem. Because other scientists may also beworking on this same problem or similar problems, the scientist must use the library to read about what others have already discovered. This is necissary to help design experiments that have already been done by others.

Hpothesis, Experimental Inference, and Analysis

This next step is to form a hypothesis, which is a simple statement that is either true or false. For example, a scientist may exept that water is necessary to keep houseplants alive. He might form the hypothesis, " water is necessary for houseplants to remain alive," which is either true or false. The perameters of the experiment would include houseplants with water, and houseplants without water. The experiment is repeated several times. In this case, the scientist could merely observe the state of the plant (dead or alive). In other experiments, the hypothesis might require that actual measurements would be made, such as the height of the plant, the number of leaves formed, etc.

Based on the outcome of the experiments and whether the hypothesis was found to be true or proven to be false, an inference is drawn,. In our example, the houseplants without water would turn brown and die, so the scientist would infer that water is necessary to keep houseplants alive. Therefore, our hypothesis that " water is necessary for houseplants to remain alive" would be supported (i.e., not proven to be incorrect). This, and other experiments which show that water is necessary for discoveries are also discussed at scientific converences.

Theory and Law

If many different experiments are performed which all support a hypothesis, then it is likely to actually be true. If most scientist have high confidence that a hypothesis to be true, then it is now considered to be a theory. Theories are hypotheses that sre assumed to be true. If a theory is tested and retested in many ways, and no evidence can be found that proves that it is not true, then it is eventually considered to be a fact of nature, and is referred to as a " law"..

Chapter 2

Cell - The Basic Unit Of Life

A cell is a microscopic, structural and functional unit of living organisms capable of independent existence. (e.g. Ameba). All living things are composed of cells. Some functioning cells come together to form a tissue and tissues collectively form organs. In more complex living organisms, organs work together for the purpose of survival as system. However, in all living organisms, the cell is a functional unit and all of biology revolves around the activity of the cell.

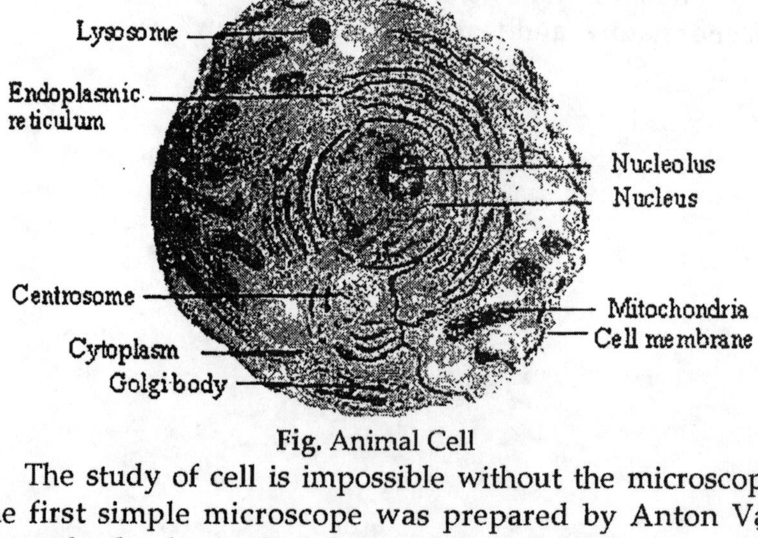

Fig. Animal Cell

The study of cell is impossible without the microscope. The first simple microscope was prepared by Anton Van Leewenhoek who studied the structure of bacteria, protozoa, spermatozoa, red blood cells etc. The word 'cell' was first coined by Robert Hooke in 1665 to designate the empty honey-comb like structures viewed in a thin section of bottle cork which he examined. He was impressed by the microscopic compartments in the cork as they reminded him of rooms in a monastery which are known as cells.

He therefore referred to the units as cells. In 1838, the German botanist Matthios Schleiden proposed that all the plants are made up of plant cells. Then in 1839, his colleague, the anatomist Theodore Schwann studied and concluded that all animals are also composed of animal cells. Schwann and Schleiden studied a wide variety of plant and animal tissues and proposed the "cell theory" in 1839. It stated that "all organisms are composed of cells." But still the real nature of a cell was in doubt. Cell theory was again rewritten by Rudolf Virchow in 1858.

In his theory he said that all living things are made up of cells and that all cells arise from pre-existing cells. It was German biologist Schulze who found in 1861 that the cells are not empty as were seen by Hooke but contain a 'stuff' of life called protoplasm. During the 1950s scientists developed the concept that all organisms may be classified as prokaryotes or eukaryotes. For example, in prokaryotic cells, there is no nucleus; eukaryotic cells have a nucleus. Another important difference between prokaryotes and eukaryotes is that the prokaryotic cell does not have any intracellular components. Bacteria and blue- green algae come under the prokaryotic group, and protozoa, fungi, animals, and plants come under the eukaryotic group.

MODERN CELL THEORY

Modern biologists have made certain additions to the original cell theory which now states that:

- All organisms are made up of cells.
- New cells are always produced from pre-existing cells.
- The cell is a structural and functional unit of all living things.
- The cell contains hereditary information which is passed on from cell to cell during cell division.
- All cells are basically the same in chemical composition and metabolic activities.

As the cells of different parts of an organism (such as epithelial cell, muscle cell, nerve cell, etc). vary in shape, size and internal structure, no cell can be described as a typical one. In spite of all these variations, they are all cells, showing certain common fundamental properties.

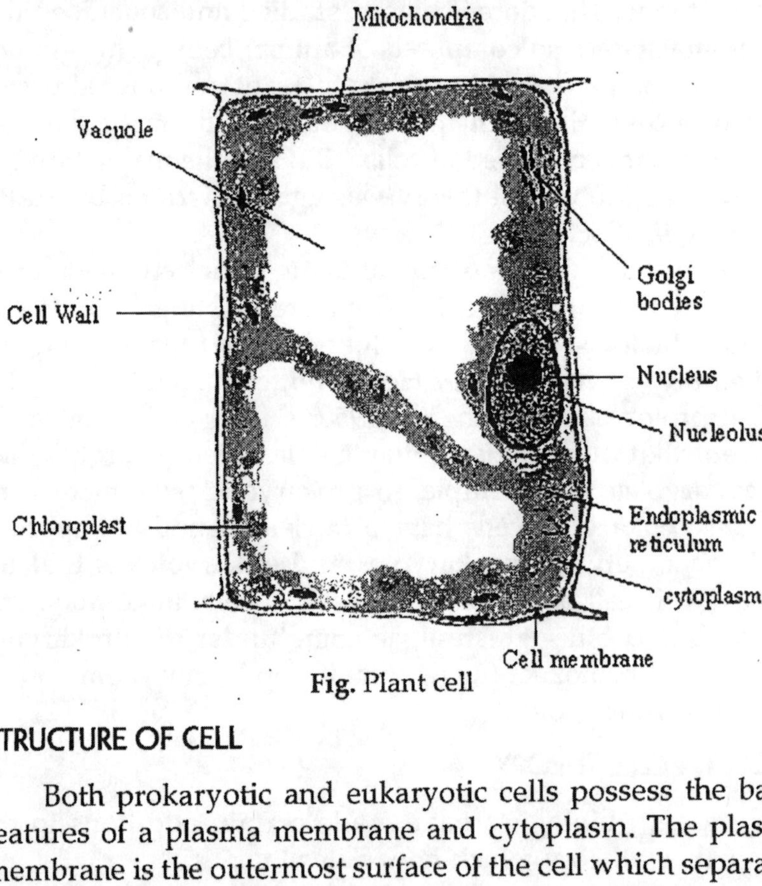

Fig. Plant cell

STRUCTURE OF CELL

Both prokaryotic and eukaryotic cells possess the basic features of a plasma membrane and cytoplasm. The plasma membrane is the outermost surface of the cell which separates the cell from the environment. The cytoplasm is the aqueous content within the plasma membrane.

Plasma Membrane

It is like any other membrane in the cell but it plays a very important function. It forms the border of a cell, so it is also called the cell membrane. It is primarily composed of proteins and phosphalipid. The phospholipids occur in two layers referred to as a bilayer. Protein is embedded within the lipid layer, or attached to the surface of it. The plasma membrane is elastic and very fluid because of protein and lipid. Normally the function of the plasma membrane is that of a gate-keeper. It allows certain important substances to enter and exit the cell.

Cytoplasm and Organelles

The cytoplasm is a semi-solid substance which is present in the cell and which gives structure, size, shape and foundation to the cell. It is enclosed by the plasma membrane. Within the cytoplasm are a number of microscopic bodies called organelles that perform various functions essential for the survival of the cell:

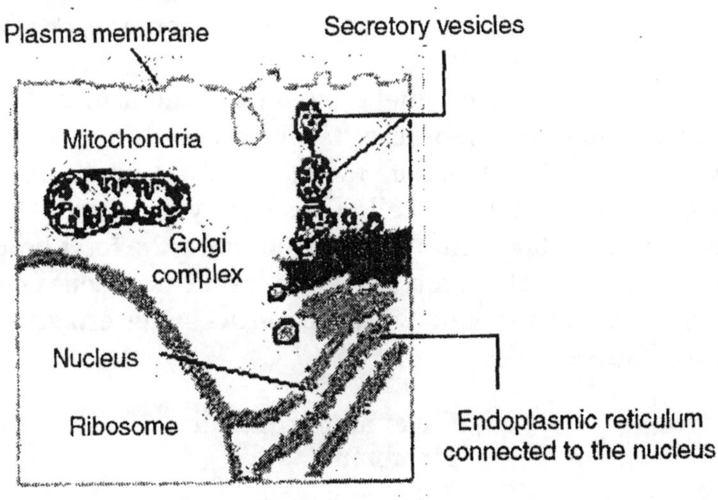

Fig. Endoplasmic Reticulum with the Nucleus and the Golgi complex

Endoplasmic Reticulum (ER)

Endoplasmic Reticulum is one of the important organelles present in the cytoplasm. Endoplasmic reticulum is a series of membranes which extend throughout the cytoplasm in eukaryotic cells. In certain cases in ER there are submicroscopic bodies called ribosomes which are involved in production of protiens.

- Rough ER: In this kind of ER the ribosomes are presenton the surface.The endoplasmic reticulum is responsible for protein synthesis in a cell. Ribosomes are suborganelles in which the amino acids are actually bound together to form proteins. There are spaces within the folds of ER membrane are known as cisternae.

- Smooth ER: This type of ER does not have ribosomes.

Another organelle is the Golgi body or Golgi apparatus (G.A.). The Golgi body is a series of flattened sacs usually curled at the edges. Proteins which have formed in ER are processed in G.A. After processing, the final product is discharged form the G.A. At this time the G.A. bulges and breaks away to form a dropline vesicle known as secretory vesicles. The vesicles move butward to the cell membrane and either insert their protien contents in the membrane, or release their contents outside the cell.

There is another organelle which is related to the Golgi apparatus called the lysosome. The lysosome is derived from the Golgi body. It is a sac of enzymes in the cytoplasm, used for digestion within the cell. These enzymes break down particles of food taken into the cell and make the food product available for use. There are also cytoplasmic organelles called peroxisomes in the cell which produces the enzymes to degrade fatmolecules.

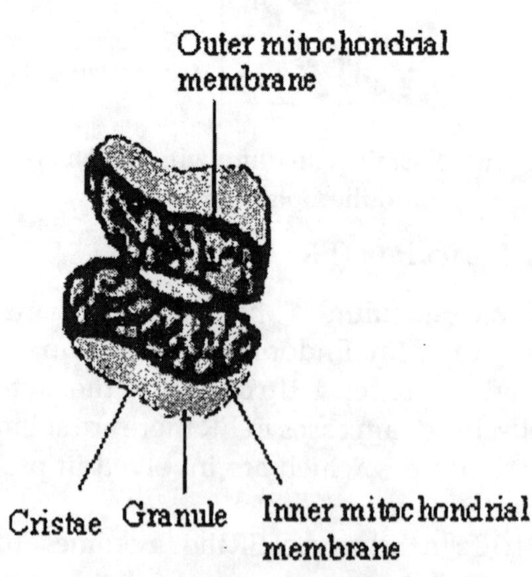

Fig. A cross-section of a Mitochondrion

Mitochondria

Is another organelle of the cell. It is called the "power house of the cell" because it stores and releases the energy of the cell. The energy released is used to form ATP (adenosine triphosphate).

Nucleus

Prokaryotic cells don't have a nucleus but eukaryotic cells have a nucleus situated in the cytoplasm. The nucleus mainly contains DNA (i.e., Deoxyribonucleic acid). DNA is organized into linear units called chromatin. Genes are the functional segments within the chromosome. Each chromosome consists of approximately 1,000,000 genes. The chromatin is coiled around nuclear protiens called histones. When chromatin is coiled, it forms chromasomes.

Genes contain the coding for all the protiens in a cell of an animal or plant. The nucleus of the cell is surrounded by an outer membrane called the nuclear envelope. The nuclear membrane resembles the plasma membrane in its function. It is also a double layer membrane consisting of two lipid layers similar to those in the plasma membrane. Pores in the membrane allow the internal nuclear part to communicate with the cytoplasm of the cell.

Prokaryotic cells don't have a nucleus but they do possess DNA which exists freely in the cytoplasm. In bacteria a single looped chromosome consists of 4,000 genes.In plant cells, organelles called chloroplasts exist. Due to chloroplasts plants look green in colour. The chloroplasts function in the process of photosynthesis.

During this process, chloroplasts convert the energy in the sunlight into energy of carbohydrate molecules. Energy from the sun comes in the form of photons i.e. a package of energy which gets converted into carbohydrate energy. Chloroplasts consist of a green pigment called chlorophyll. Because chlorophyll molecules absorb most wavelenghths of light except green, they reflect green light and appear green to our eyes. chlorophyll is normally present in that area of a plant where sunlight can reach easily. For example, the leaves and stem of a plant are green. On the contrary, the roots don't have chloroplast so they are not green.

Cytoskeleton

Cytoskeleton is an interconnecting system of fibers and threads and interwoven molecules that give structure to the cell. The main component of the cytoskeleton are microtubules, microfilaments and intermediate filaments. They are all made up of proteins.

Centriole

Centriole is another organelle present in the cell. It is cylindrical in shape and always occurs in pairs. Centrioles are involved in cell division.

Vacuole

Another organelle seen in the plant cell is the vacuole. The vacuole forms about 75% of the plant cell. In the vacuole, the plant stores nutrients as well as toxic wastes. If pressure increases within the vacuole it can increase the size of the cell. In this case the cell will become swollen. If the pressure increases further the cell will get destroyed.

Many cells have structures attached to them called Flagella or Cilia. Flagella are seen in the single-celled plant and protozoans, and cilia are commonly seen in animal cells. Flagella are long hairlike extensions that extend from the cell and help in locomotion. Animal sperm has flagella which permit locomotion. Cilia are shorter and more numerous than flagella. Rows of cilia move in waves to move the cell (prokaryotes like paramecia), or to move fluids around the cell (e.g., respiratory epithelial cells).These cells help in the removal of particles from the tract.

Cell Wall

Plant cells possess a cell wall. It is a structure which is present outside the cell membrane. It is not very thick. In bacteria the cell wall is very thick and rigid: this gives shape to the bacteria. In a eukaryote cell, the cell wall is not identical in different animals. In fungi, the cell wall is made up of chitin which is a polysaccharide. In plant cell there is no chitin. Cell walls are composed of another polysaccharide called cellulose.

The cell wall provides support to the structure to the cell. It also saves the cell from mechanical pressure.; it is not a selective (semipermeable) membrane like the plasma membrane. When bacteria enter the human body the cell wall this is recognized as a foreign substance in the body; this is how our immune system recognizes and destroys bacteria. Prokaryotic cells don't have a nucleus but they do possess DNA which exists freely in the cytoplasm. In bacteria a single looped chromosome consists of 4,000 genes.In plant cells, organelles called chloroplasts exist. Due to chloroplasts plants look green in colour. The chloroplasts function in the process of photosynthesis.

During this process, chloroplasts convert the energy in the sunlight into energy of carbohydrate molecules. Energy from the sun comes in the form of photons i.e. a package of energy which gets converted into carbohydrate energy. Chloroplasts consist of a green pigment called chlorophyll. Because chlorophyll molecules absorb most wavelenghths of light except green, they reflect green light and appear green to our eyes. chlorophyll is normally present in that area of a plant where sunlight can reach easily. For example, the leaves and stem of a plant are green. On the contrary, the roots don't have chloroplast so they are not green.

Cytoskeleton

Cytoskeleton is an interconnecting system of fibers and threads and interwoven molecules that give structure to the cell. The main component of the cytoskeleton is microtubules, microfilaments and intermediate filaments. They are all made up of proteins.

Centriole

Centriole is another organelle present in the cell. It is cylindrical in shape and always occurs in pairs. Centrioles are involved in cell division.

Vacuole

Another organelle seen in the plant cell is the vacuole. The vacuole forms about 75% of the plant cell. In the vacuole, the plant stores nutrients as well as toxic wastes. If pressure increases within the vacuole it can increase the size of the cell. In this case the cell will become swollen. If the pressure increases further the cell will get destroyed.

Many cells have structures attached to them called Flagella or Cilia. Flagella are seen in the single-celled plant and protozoans, and cilia are commonly seen in animal cells. Flagella are long hairlike extensions that extend from the cell and help in locomotion. Animal sperm has flagella which permit locomotion. Cilia are shorter and more numerous than flagella. Rows of cilia move in waves to move the cell (prokaryotes like paramecia), or to move fluids around the cell (e.g., respiratory epithelial cells).These cells help in the removal of particles from the tract.

Cell Wall

Plant cells possess a cell wall. It is a structure which is present outside the cell membrane. It is not very thick. In bacteria the cell wall is very thick and rigid: this gives shape to the bacteria. In a eukaryote cell, the cell wall is not identical in different animals. In fungi, the cell wall is made up of chitin which is a polysaccharide. In plant cell there is no chitin. Cell walls are composed of another polysaccharide called cellulose.

The cell wall provides support to the structure to the cell. It also saves the cell from mechanical pressure.; it is not a selective (semipermeable) membrane like the plasma membrane. When bacteria enter the human body the cell wall this is recognized as a foreign substance in the body; this is how our immune system recognizes and destroys bacteria.

MOVEMENT THROUGH THE PLASMA MEMBRANE

The cell membrane separates the cell from the external environment. In order to communicate with the external environment for the purpose of survival (e.g., for the consumption of food, minerals etc.), there is movement in the cytoplasm and plasma membrane. This movement occurs through several mechanisms which are listed below:

Osmosis

One method of movement through the membrane is osmosis. Osmosis is the movement of water. Osmosis often occurs across a membrane that is semipermeable. A semipermeable membrane allows only certain molecules to pass through while keeping other molecules out. Osmosis is really a type of diffusion involving only water molecules.

Diffusion

Another method of movement through the membrane is diffusion. Diffusion is the movement of molecules from a region of higher concentration to one of lower concentration. This movement occurs due to molecules which constantly collide with each other. The total effectual momentum of the molecules is away from the region of high concentration to the region of low concentration. Diffusion is the random movement of molecules. The exchange of molecules (taking place from a higher concentration region to a lower concentration region) leads to the formation of concentration gradient. Diffusion phenomena can be seen by letting a drop of dye into water. The colour of the dye gets diffused throughout the water.

Facilitated diffusion

A third method is facilitated diffusion which occurs across the plasma membrane. This type of diffusion is very specialized. This occurs only in cases where specific proteins in the membrane permit only certain molecules across the membrane. These membrane proteins allow movement in the direction that diffusion would normally take from a region with a higher concentration of molecules to a region with a lower concentration of molecules. No energy use is required for facilitated diffusion.

Active transport

A fourth method for movement across the membrane is active

transport. When active transport takes place, a protein moves a certain substance across the membrane, usually from a region of lower concentration to a region of higher concentration. As we know this movement is against the concentration gradient, hence energy is required for this movement.

Normally the cell gets its energy from ATP (adenosine triphosphate). For example, in cardiac muscles, active transport takes place. In these cells, sodium ions are constantly transported out of the cell. The cellular compartment is a region of high concentration of sodium ions. Buildup of electrically charged ions allows changes in voltage over the cell membrane which affects contraction of muscle cells.

Endocytosis

This another mechanism of movement across the plasma membrane. In this type, a small patch of plasma membrane encloses particles or tiny volumes of fluid which are at or near the cell surface. The membrane enclosure then sinks into the cytoplasm and breaks off from the membrane, forming a vesicle that moves into the cytoplasm. When the vesicle contains particulate matter, the process is called phagocytosis. When the vesicle contains liquids or droplets of fluids the process is called pinocytosis.

THE CHEMICAL BASIS OF LIFE

Biology is the study of the natural world - the living world around us. Formerly the biologist merely classified plants and animals according to their size, shape (i.e. anatomy), their physiology and their existence in nature. Scientists later discovered the chemical and physical bases of living things. From their own study they realised that there is a chemical similarity between a plant cell and an animal cell. This leads us to the study of the chemical basis of life.

STRUCTURAL ORGANIZATION AND THE CHEMICAL BASIS OF LIFE

The universe and living bodies are composed of matter which occupies space and possesses mass. Matter can exist in four forms -solid, liquid, gaseous and plasma. Matter is made up of basic substances called elements. There are over a hundred elements recognized by scientists today. An element is made up of identicle atoms. Calcium, Carbon, hydrogen, iron, sodium, etc. are some elements.

Atoms

Each element is made up of one particular kind of atom. Atoms are the smallest part of an element which do not share the properties of the element. An atom consists of 3 basic particles:

- Protons, which are positively charged and are present in the nucleus.
- Electrons, which are negatively charged and which rotate around the nucleus.
- Neutrons, which are present in the nucleus and which don't have any electricle charge. A neutron has approximately the same mass as a proton.

In the nucleus of the atom the proton and neutron are firmly attached to each other. The chemistry of the atom is dependent upon the number of protons and electrons in the atom. Protons and electrons are present in the same number in the atom, which leads to a neutral electrical charge in the atom. The nucleus of an atom is surrounded by shells or orbits of electrons. In the first orbit there are a maximum of 2 electrons. From the second shell to the last shell, every shell has a maximum of 8 electrons. When the final orbit of an atom is complete with electrons it becomes a stable atom.

When there is a single electron or more missing in the last orbit of the atom, the atom becomes active allowing for a chemical reaction with another atom. During the reaction there is sharing or exchange of electrons. Atoms are most stable when they have a full outer shell. If they can, they will either share one or more electrons with another atom so that both atoms " think " they have full outer shells, or sometimes an atom that " needs " one or two electrons to have a full outer shell will actually take electrons another atom that has only one or two electrons in its outer shell; therefore, both atoms end up with full outer shells.

Molecules

In biology we study molecules as part of molecular biology, molecular interaction, etc. Atoms combine chemically in a specific order to form molecules. For instance, two atoms of hydrogen combine with one atom of oxygen to form a single molecule of water. A molecule is the smallest particle of a substance existing freely yet retaining the characteristics of that substance. A collection of molecules forms a compound. Properties of the compound depend upon molecules and atoms present in the molecule. Consider this example: the molecular weight of the compound, water, is 18. This weight has been calculated taking into account the weight of its molecular components H and O.

$$H_2O \rightarrow H + H + O \rightarrow 1 + 1 + 16 \rightarrow 18$$

Some molecules are formed from atoms of the same element e.g. oxygen molecule (O_2) is formed from 2 atoms of oxygen; ozone (O_3) is formed from 3 atoms of oxygen. But, oxygen or ozone cannot be called a compound because in a compound we require atoms of different elements.

When the compound is formed it contains different elements. These elements stay together by means of links between them. In scientific terminology this link is called a bond. There are various kinds of bonds. In a compound, when one atom of an element gives away an electron, this exchange creates a bond between these two elements. This is an ionic bond. This bond consists of an electromagnetic force which is formed due to exchange of charge [electron]; the atoms conducting the exchange are called ions. This eletromagnetic force attracts two opposite charges: a positive charge in the atom which gives away electrons and a negative charge in the atom which takes up electrons. This attraction force forms the ionic bond.

A second important type of bond is called the covalent bond. Such a bond is formed when two atoms share one or more electrons with one another. For example, in water (H_2O) 2 atoms of hydrogen share 2 electrons with oxygen. In ammonia 3 hydrogen atoms share one electron each with a single nitrogen atom. This leads to the formation of ammonia

(NH_3). Between atoms when one pair of electrons is shared, a single covalent bond is formed. When two pairs of electrons are shared a double covalent bond is formed.

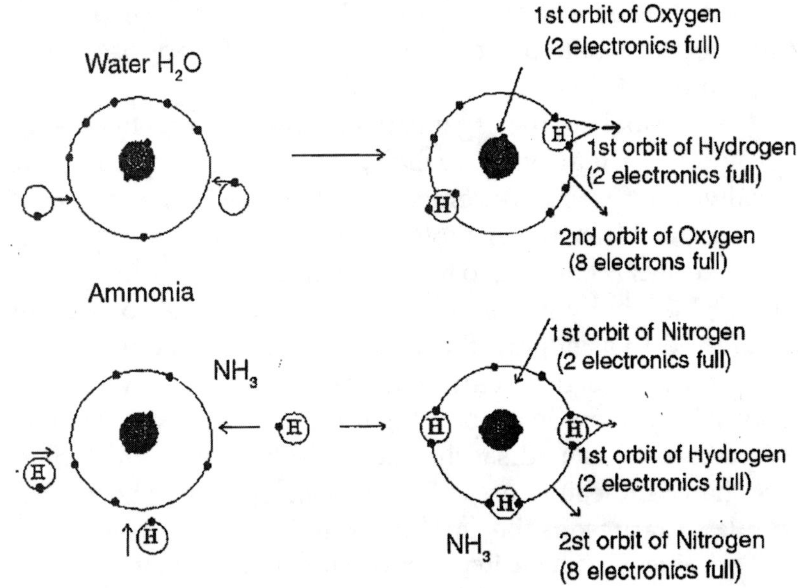

Acids and Bases or Alkalis

If a compound reacts with water and releases hydrogen ions (H^+) ions then this compound is called an acid, and is said to be acidic in nature. For example, when hydrogen sulphide is mixed with water it releases hydrogen ions and the solution becomes one of sulphuric acid. Other chemical compounds when dissolved in water attract hydrogen atoms. These substances are called bases or alkalis. For example, when sodium hydroxide (NaOH) is mixed with water, sodium hydroxide attracts hydrogen ions from H_2O, and (OH^-) ions remain. So these substances that remove (H^+) from water act as bases or alkalis.

ORGANIC COMPOUNDS

Many chemical compounds in living organisms are known as organic compounds' which contain C,H and O. We have seen that an organism is formed primarily from six elements: C, H, O, N, P, Ca. The study of organic compounds is called organic chemistry.

CARBOHYDRATES

There are plenty of organic compounds present in nature. All living things contain basically four types of organic compounds.Carbohydrates form the first category of organic compounds.

For metabolism the organism requires energy. This energy is provided primarily by carbohydrates. Carbohydrates are basically composed of 3 elements, C, H, and O. The ratio of H to O is 2:1, as in a water molecule (H_2O). There are types of carbohydrates according to the complexity of the carbohydrate molecule. Carbohydrates and usually taste sweet to humans are referred to as sugar. If a carbohydrate is made up from a single molecule it is called monosaccharide. When the carbohydrate is made up of 2 sugar molecules linked together it is referred to as a adisaccharide Carbohydrates which have more than 3 molecules are called polysaccharides. The general formula to represent the carbohydrate is $Cx(H_2O)y$.

Table: Schematic Representation of Carbohydrates

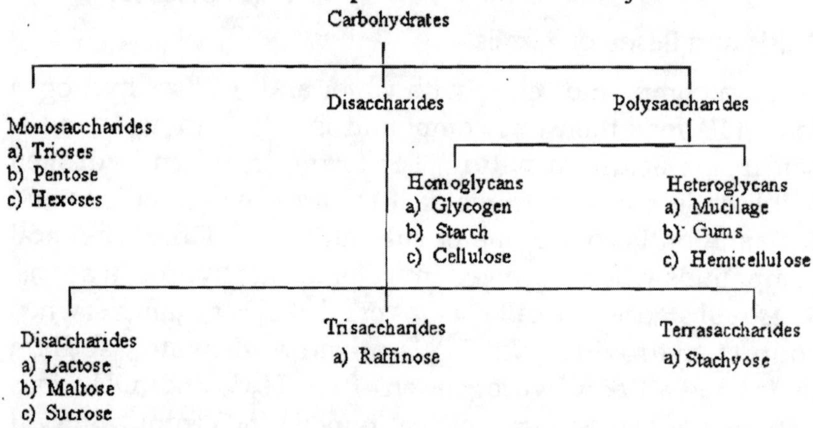

Monosaccharides

They are the simplest soluble sugar. Depending on the number of carbon atoms present, monosaccharides are further classified as:

- Trioses → (3 carbons) $C_3H_6O_3$ e.g. glyceraldehyde
- Pentoses → (5 carbons) $C_5H_{10}O_5$ e.g. ribose and deoxyribose
- Hexoses → (6 carbons) $C_6H_{12}O_6$ e.g. glucose

Glucose $C_6H_{12}O_6$ is a basic form of fuel in all living things. It is soluble in blood plasma and water and so it is transported by body fluids to all cells in the body. In cells it is metabolized and releases energy. Glucose is also the main product of photosynthesis and also an initiating material for cellular respiration.

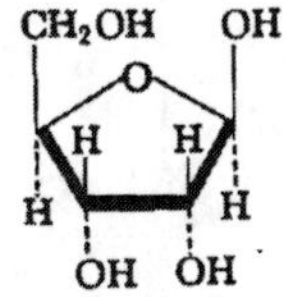

Fig. A Molecular Representation of Glucose

Disaccharides

These carbohydrates contain two monosaccharides linked together and accordingly they are known as:

- Disaccharide: contains two monosaccharides e.g. lactose, maltose, sucrose

  Maltose → Glucose + Glucose

  Sucrose → Glucose + Fructose

  Lactose → Glucose + Galactose

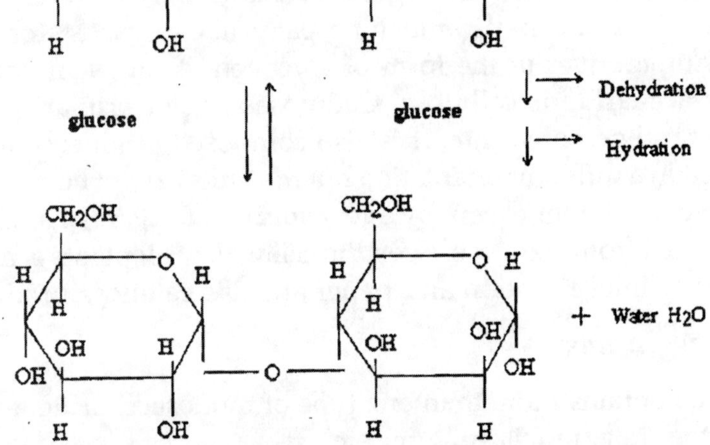

- Trisaccharide: Containing 3 Monosaccharides. e.g. Raffinose
- Tetrasaccharide: Containing 4 Monosaccharide e.g Stachyose

Polysaccharides

General formula n $(C_6H_{10}O_5)$. These complex carbohydrates are formed by chains of at least ten monosaccharides. They are of two types:

Homoglycans

Containing only one type of monosaccharide (e.g. glycogen, starch, cellulose, contain only glucose molecules). Starch is a very important polysaccharide because it is formed through a chain of hundreds or thousands of glucose units. Carbohydrates in plants are stored in the form of starches. Starch contained in energy rich food like rice, corn, and potatoes form part of the staple diet of most people.

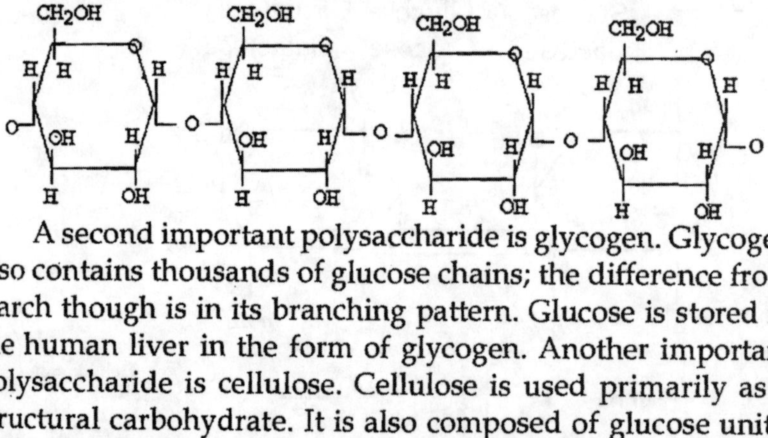

A second important polysaccharide is glycogen. Glycogen also contains thousands of glucose chains; the difference from starch though is in its branching pattern. Glucose is stored in the human liver in the form of glycogen. Another important polysaccharide is cellulose. Cellulose is used primarily as a structural carbohydrate. It is also composed of glucose units, linked in a different orientation but the units cannot be released from one another except by a few species of organisms. Wood is formed from cellulose. Even the cell wall of all plants is made up of cellulose. Cotton and paper are also cellulose products.

Heteroglucans

It contains more than one type of monosaccharide linked together (e.g. mucilage, gum etc.)

PROTEINS AND ITS DERIVATIVES

Proteins are the fundamental chemical compounds of the protoplasm indispensable for vital life processes. They are complex, large molecules each containing thousands of atoms. proteins contain nitrogen in addition to carbon, hydrogen and oxygen; they usually also contain phosphorus and sulfur. These compounds are polymers of unit structures called amino acids, represented chemically as:

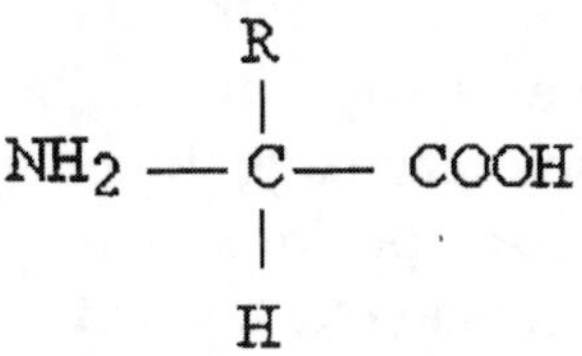

Fig. Amino acid

NH_2 is an amino group, - COOH is the carboxyl group,and R represents the variable chain forming different amino acids. There are 20 diferent of amino acids. The amino acids differ depending on the nature of the R group. Examples of. amino acids are valine, alanine, glutamic acid, tyrosine and histidine.

Two molecules of amino acids are joined by the carboxyl group of one ammino acid with the amino group of the other by loss of one molecule of water. This process is called dehydration synthesis and the bond thus formed between two molecules is referred to as the peptide or peptide bond.

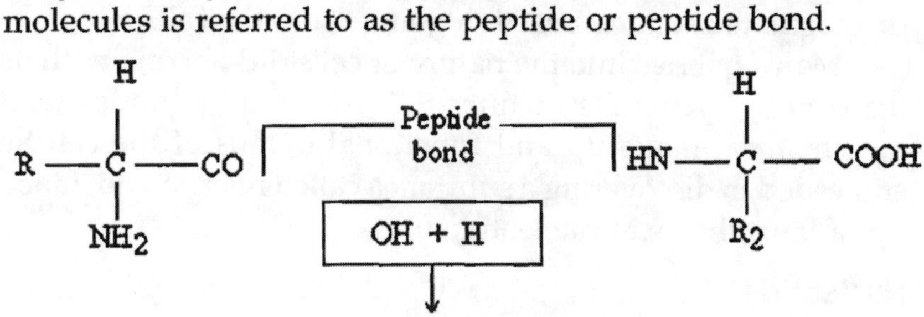

Fig. Peptide Bond

There are 3 types of proteins namely:

- Simple proteins: like albumins and globulins formed by group of amino acids only.
- Derived proteins: like proteose and peptones which are hydrolytic cleavage products of complex proteins.

- Conjugated proteins: like nucleo proteins (Proteins + nucleic acid), lipoproteins (protein + lipid), or glycoproteins (protein + carbohydrates) which are formed by the combination of proteins with some non-protein molecule. This non-protein portion is called Prosthetic group.

All living things require protein for survival. In fact an organism is constructed by means of proteins. All living things then, in any form - liquid, solid, or plasma - contain proteins. Protein is also seen as a supporting tissue with main tissue. Bone, tendons, muscle, cartilage, ligaments are all formed of protein. Enzymes are a specified class of proteins. Enzymes act as catalysts in chemical reactions of the body. They are not used up by the reaction, rather they remain chemically unchanged and available to catalyze succeeding reactions.

NUCLEOTIDES AND NUCLEIC ACID

Every organism reproduces within its life span. This is accomplished through cell divisions and is regulated by many kinds of protiens. The information forsynthesizing unique proteins is located in the nucleus of the cell. It is called the genetic code which is the "blue print" for producing specific sequences of the amino acids in proteins. Thus the genetic code can regulate chemical reaction going on in the cell.

Man's queries into the nature of cells did not end with its discovery of general structures. In an attempt to understand the chemical make up and functional details of the cell he succeeded in discovering a substance called nucleic acid, made up of long chains of nucleotide units.

Nucleotide

It is the structural unit of nucleic acid. Each nucleotide is composed of:

- Pentose sugar
- Phosphate group
- One of four nitrogen bases attached to the pentose sugar. A nucleotide without a phosphate group is a nucleoside.

Nucleic acids

They are complex, large biomolecules formed of many units called nucleotides.

Nucleic acids are of two types:

- DNA - Deoxyribonucleic acid and, (2) RNA - Ribonucleic acid

The DNA of cells contains genetic information in a coded form, and is only present in the nucleus of the cellis formed from DNA, plus a few special organelles. RNA and is present in the cytoplasm and in the nucleus of the cell. DNA and RNA differ from one another in their components. DNA contains the pentose sugar, deoxyribose, while RNA contains ribose. There are also small differences in the types of nitrogen bases found in DNA and RNA.

LIPIDS

Fats and their derivatives are collectively called lipids. Fats are compounds containing fatty acids and glycerol. They are composed of carbon, hydrogen and oxygen, but less oxygen than that in carbohydrates. Fatty acids are long chains ofCH_2 groups with terminal methyl and carboxyl groups with general formula $CH_3[CH_2]n\text{-}COOH$, while glycerol molecule contains a chain of three carbon atoms and has a formula $C_3H_6O_3$. In the formation of fat, three molecules of fatty acids are combined with three-OH groups on one molecule of glycerol, with removal of 3 molecules of water which is represented as follows:

glycerol + 3 fatty acids ⇌ Fat + 3 water molecules

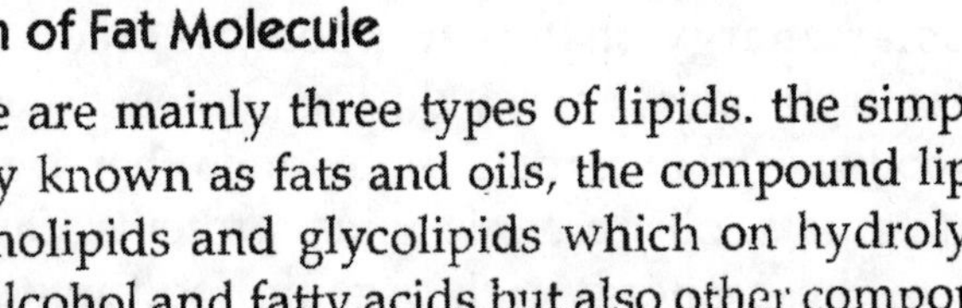

Formation of Fat Molecule

There are mainly three types of lipids. the simple lipids, commonly known as fats and oils, the compound lipids such as phospholipids and glycolipids which on hydrolysis yield not only alcohol and fatty acids but also other compounds and derived lipids such as steroids which include cholesterol, Vitamin D, estrogen, testosterone, cortisol, etc. Lipids are practically insoluble in water but are soluble in organic solvents like chloroform, ether and benzene.

Fats stored in cells are usually clear oil droplets called globules. Because fats do not dissolve in water, animals store fat in large clear globules in the cells of adipose tissue. The enzyme lipose breaks down fats into fatty acids and glycerol which can be furthur broken down to produce energy.

CELLULAR RESPIRATION

All forms of life have one basic common requirement, and that is energy. Energy gives the capacity to do work. Organisms need energy for existence and maintenance of life. The cell is the basic structural and functional unit of an organism and it performs various metabolic and other vital activities. These activities need energy, i.e. biologically usable forms of chemical energy, such as the energy-rich organic compound called ATP (adenosine tri-phosphate). ATP molecules act as energy mediators in cells. For example, ATP undergoes a breakdown (hydrolysis) and provides energy for energy requiring reactions. On the other hand, ATP is synthesized by input of energy made available through some energy-yielding reactions in the cell.

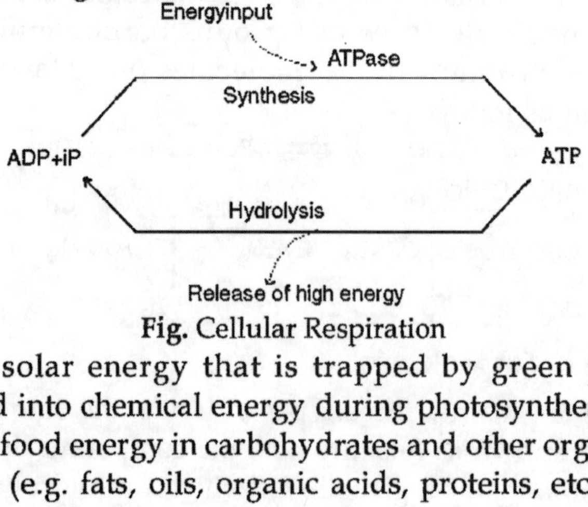

Fig. Cellular Respiration

The solar energy that is trapped by green plants is converted into chemical energy during photosynthesis and is stored as food energy in carbohydrates and other organic food materials (e.g. fats, oils, organic acids, proteins, etc.). In cell metabolism, these energy-rich organic food materials are oxidized and the stored energy is released as free energy. This is utilized by the cell to perform various cellular activities. Part of the energy is lost as heat. In the cells, the process which supplies energy through oxidation of organic compounds is called cellular respiration.

RESPIRATION

"Respiration is an intracellular process of oxidation-reduction reactions in which the complex organic food materials are broken down in a step-wise manner to form simpler end products, with the release of energy and carbon dioxide."

It is basically an energy releasing and supplying process. The energy released in respiration is of two types:

- Chemical energy, i.e. ATP. It is utilized for the cellular activities.
- Heat energy. It is mostly lost. Depending upon the availability of oxygen as an oxidant, respiration is of two types:
  - Aerobic respiration, in which oxygen in used in the respiratory break down of organic substrate;
  - Anaerobic respiration, in which oxygen is not used in the respiratory breakdown of organic substrate.

ULTRASTRUCTURE AND FUNCTIONS OF MITOCHONDRION

Mitochondria are the special protoplasmic organelles distributed in the cytoplasm of eukaryotic cells. Mitochondria contain the biochemical machinery involved in cellular respiration which take energy from breakdown of glucose and produce energy-rich ATP molecule which fuel the biochemecal reactions in the rest of the cell. Hence, mitochondria are described as the 'power houses' of cells. Mitochondria were first observed by Altmann. Hogeboom discovered that mitochondria are the site of aerobic respiration. Ultrastructure Mitochondria are microscopic and granular or cylindrical, and are bound by two lipo-protein membranes called outer and inner membranes.

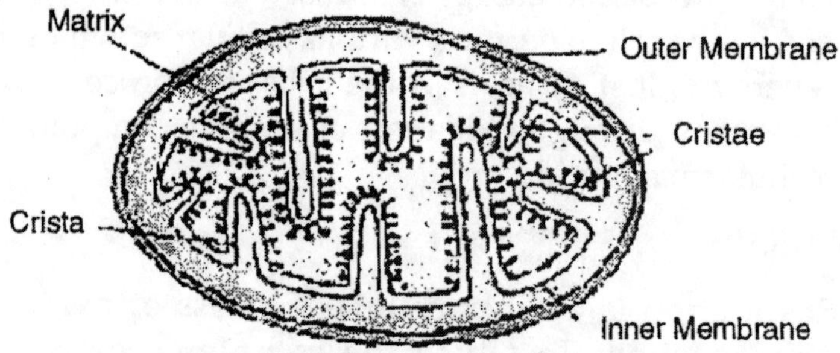

Fig. A Mitochondrion in Section.

Ultrastructure of a Mitochondrion

These are separated from each other by the inter-membrane space. The outer membrane is smooth, while the inner membrane is highly folded. The folds are called cristae, and project into the inner space of the mitochondrion and appear finger-shaped in the section. The internal space of the mitochondrion (enclosed within the folded inner membrane) is filled with dense proteinaceous material and is called the matrix. It contains small ribosomes and mitochondrial DNA (DNA) invited in the replication of mitochondria. The matrix also contains all the enzymes required for the Kreb's cycle reactions during the second phase of aerobic respiration.

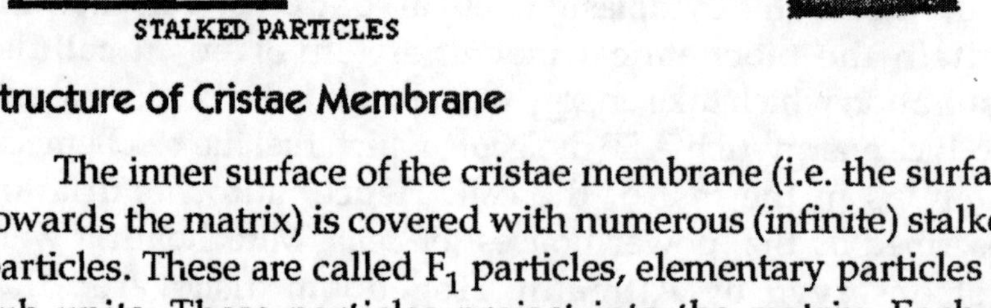

Structure of Cristae Membrane

The inner surface of the cristae membrane (i.e. the surface towards the matrix) is covered with numerous (infinite) stalked particles. These are called F_1 particles, elementary particles or sub units. These particles project into the matrix. Each F_1 particle has three parts, viz., the head piece, the stalk and the base piece. The respiratory chain is situated in the cristae membrane where the F_1 particles are present. The chain consists of enzymes and co-enzymes which form the electron transport system (ETS) in the mitochondrion. These enzymes and co-enzymes of the ETS act as the electron acceptors in the aerobic respiration reactions oxidative phosphorylation).

Functions of Mitochondria

- These are the sites of the phase-II reactions of aerobic respiration.
- Kreb's cycle reaction takes place in the matrix.
- Terminal oxidation (utilization of O_2 and formation of water) and oxidative phosphorylation (formation of ATP) take place in cristae membrane.
- ATP molecules formed in aerobic respiration are stored in mitochondria and are supplied for cellular activities. Hence, mitochondria are the energy supplying power houses of cells.

AN OVERVIEW OF CELLULAR RESPIRATION

Respiration is of two types: aerobic and anaerobic respiration.

Aerobic Respiration

"When free or molecular oxygen participates in the respiratory breakdown of organic substrate, it is called aerobic respiration." (In this process, free molecular oxygen acts as the final electron acceptor.). Aerobic respiration is represented by the following chemical equation.

$$C_6H_{12}O_6 + 6O_2 \rightarrow 6CO_2 + 6H_2O + \text{Energy}$$

Energy → Heat energy = 673 K.Cal/glucose molecule
→ Chemical Energy = 38 ATP / glucose molecule

Anaerobic Respiration

When the respiratory breakdown of organic substrate takes place without participation of free molecular oxygen, it is called anaerobic respiration." (In this process, free or molecular oxygen is not the final electron acceptor.). Anaerobic respiration is represented by the following chemical equation.

$$C_6H_{12}O_6 \rightarrow 2C_2H_5OH + 2CO_2 + \text{Energy}$$

Energy → Heat energy = 28 K.cal/glucose molecule

Energy → Chemical Energy = 2 ATP / glucose molecule.

The process of respiration is completed in two phases. Reactions in Phase I are called glycolysis and are common to both types of respiration. Reactions in Phase II depend upon whether O_2 is utilized or not.

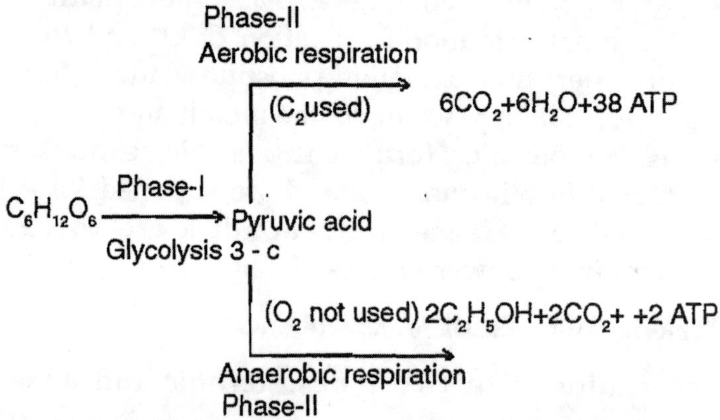

GLYCOLYSIS (E.M.P. PATHWAY = EMBDEN - MEYERHOF - PARNAS PATHWAY)

Important Features

- Glycolysis represents phase I of respiration.
- This pathway is common to both aerobic and anaerobic respiration.
- These reactions are completed in the cell cytoplasm (cytosol). Therefore, glycolysis is also called cytoplasmic respiration.
- Glycolysis involves a series of reactions in which a molecule of 6-C glucose is broken down in a step-wise manner into two molecules of 3-C pyruvic acid.
- Each step (reaction) is controlled by a specific enzyme and is reversible.
- Reactions in the glycolysis are anaerobic in nature,(i.e. they do not require molecular oxygen). 6-C Glucose is the most commonly used organic substrate for respiratory breakdown because
- It is most readily available.

- It is soluble and can be easily transported and supplied..

Reactions of glycolysis (EMP pathway) involve:

- Phosphorylation of glucose (preparatory phase), isomerization to fructose-6-phosphate, and phosphorylation to fructose-I, 6-biphosphate.
- Cleavage of Fructose 1, 6 diphosphate to 2 3-C PGAL.
- Formation of 3-C pyruvic acid (end product of glycolysis). Various steps (intermediate reactions) in glycolysis are as follows.

Phosphorylation of Glucose

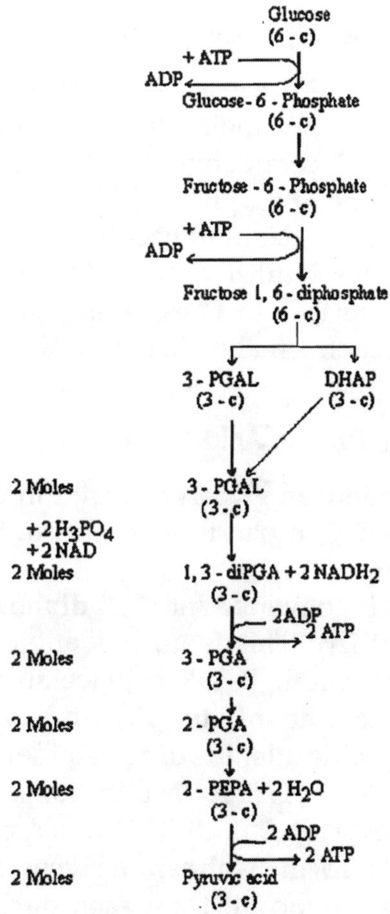

Fig. Glycolysis (EMP pathway). Various steps in glycolysis

The initial steps in glycolysis involve the phosphorylation of glucose into an unstable fructose 1, 6-diphosphate for easy breakdown. This is called the preparatory phase of glycolysis during which 2 ATP are used. The reactions are as follows.

- In the first step, 6-C glucose is converted into glucose-6-phosphate. One ATP is used in the reaction.
- The next step involves isomerization of glucose 6-phosphate into Fructose 6-phosphate.
- Fructose-6-phosphate is then phosphorylated to fructose-1, 6-diphosphate. This reaction utilizes one ATP. Thus, two ATP are used up in the preparatory phase.

Cleavage of Fructose-1, 6–diphosphate

The molecule of 6-C fructose-1, 6-diphosphate undergoes cleavage to produce two molecules of 3-C triose phosphates. One is 3-phosphoglyceraldehyde (3-PGAL) and the other is dihydroxy acetone phosphate (DHAP). The two triose phosphates can isomerize into each other. However, further reactions in glycolysis utilize only PGAL. Therefore, DHAP is first converted into PGAL. Thus, in all, two molecules of 3-PGAL are formed from the cleavage of one fructose-1, 6-diphosphate.

Formation of 3-C Pyruvic Acid

Further reactions in glycolysis result in the conversion of 3-C PGAL into 3-C pyruvic acid through the following intermediate steps:

- 3-PGAL is converted into 1, 3 diphosphoglyceric acid (1, 3 diPGA). This is an oxidation and phosphorylation reaction. It takes place in the presence of inorganic phosphate (P_i) and co-enzyme NAD (nicotinamide adenine dinucleotide). The reduced co-enzyme $NADH_2$ is formed in the process.
- In the next step, 1, 3-diPGA undergoes dephosphorylation to form 3-phosphoglyceric acid (3-PGA). Phosphate removed from each diPGA is transferred to by ADP to form ATP. Two ATP molecules are formed at this step.

- 3-PGA is transformed into 2-PGA.
- In the next step, 2-PGA is converted into 2-phosphoenol pyruvic acid (2-PEP) with the loss of water (dehydration).
- In the final step of glycolysis, 2-PEP is dephosphorylated to form 3-C pyruvic acid, which is the end product of glycolysis. The reaction produces 2 ATP.

Thus, when a molecule of 6-C glucose undergoes the reactions in glycolysis (EMP pathway), the overall process may be represented as follows.

$$\underset{6\text{-c}}{\text{Glucose}} \xrightarrow{\text{Glycolysis}} 2\ \underset{3\text{-c}}{\text{Pyruvic acid}} + 2\ NADH_2 + 4\ ATP$$

However, 2ATP are used up in the preparatory phase of glycolysis. Therefore the net gains of glycolysis are:

$$\underset{6\text{-c}}{\text{Glucose}} \xrightarrow{\text{Glycolysis}} 2\ \underset{3\text{-c}}{\text{Pyruvic acid}} + 2\ NADH_2 + 2\ ATP$$

PHASE-II OF AEROBIC RESPIRATION

Phase-II of aerobic respiration comprises a series of enzymic reactions in which 3-c pyruvic acid (the end product of glycolysis) is broken down step-wise to form the end products of aerobic respiration, (i.e. CO_2, H_2O and ATP energy), inside the mitochondria.

Phase-II involves the following two main sets of reactions:

- The T.C.A. cycle
- Terminal oxidation and oxidative phosphorylation (E.T.S. and the redox reactions).

THE T.C.A. CYCLE (TRICARBOXYLIC ACID CYCLE)

The T.C.A. (Tricarboxylic acid) cycle represents the major set of reactions during Phase-II of the aerobic respiration. It is also called the citric acid cycle or Kreb's cycle. These reactions involve the final breakdown of pyruvic acid to form the end products of aerobic respiration. The T.C.A. cycle reactions take place in the matrix of the mitochondria. Pyruvic acid is first converted into a 2-C compound called acetyl co-enzyme-A (Acetyl Co-A).

Formation of 2-C Acetyl Co-A

This is the initial step in Phase II of aerobic respiration. The reaction involves oxidation and decarboxylation of 3-C pyruvic acid to form 2-C acetyl Co-A. It in the cell cytoplasm and requires the presence of co-enzyme-A (Co-A) and NAD. It is catalyzed by complex enzymes. In this process, carbon dioxide is released and the reduced co-enzyme $NADH_2$ is formed.

$$\underset{\text{acid}}{\text{2 Pyruvic}} + \text{2Co-A} + \text{2NAD} \xrightarrow[\text{complex)}]{\text{(enzyme}} \underset{\text{(2-c)}}{\text{2Acetyl Co-A}} + 2CO_2 + 2NADH_2$$

The 2-C acetyl Co-A formed in the cytoplasm then enters the mitochondria and takes part in the Kreb's cycle reactions (the TCA cycle). Thus, acetyl Co-A is the connecting link between glycolysis (in the cytoplasm) and Kreb's cycle (in mitochondria).

T.C.A. Cycle (Kreb's Cycle)

The Kreb's cycle reactions occur in the matrix of mitochondria. All enzymes required for these reactions are present in the matrix. Each step is controlled by a specific enzyme and is reversible.

The various steps in the T.C.A. Cycle are as follows:

- Each molecule of 2-C acetyl Co-A that enters the T.C.A. cycle first combines with 4-C oxalo acetic acid (OAA), and a 6-C Citric acid is formed. One molecule of H_2O is used in the reaction.
  Thus, the first product in Kreb's cycle is citric acid, hence it is also called the citric acid cycle. As citric acid is a tricarboxylic acid (with 3-COOH groups), it is also called tricarboxylic acid (TCA) cycle.
- In the next step, 6-C citric acid is first converted into 6-C -aconitic acid (deletion of H_2O) and then into 6-C isocitric acid (addition of H_2O).
- The next reaction involves the oxidation of isocitric

acid (by removal of hydrogen) to form 6-C oxalo succinic acid. $NADH_2$ is formed in the process.

- Decarboxylation of 6-c Oxalo succinic acid results in the formation of 5-c a -Ketoglutaric acid with the liberation of carbon dioxide.
- a-Ketoglutaric acid (5-C) then undergoes oxidation (by removal of hydrogen) and decarboxylation to form 4-C succinyl Co-A. The reaction is highly complicated and takes place in the presence of Co-A and NAD. $NADH_2$ is formed and carbon dioxide is released.
- 4-C Succinyl Co-A is hydrolyzed to succinic acid (4-C) in the next step. One molecule of H_2O is used and Co-A is regenerated. The reaction is exergonic. Energy released is used for the formation of GTP (guanosine triphosphate) from GDP and inorganic phosphate.

  Subsequently, ATP is formed when GTP reacts with ADP

  GTP + ADP $\rightarrow$ ATP + GDP

  Thus, there is the direct formation of one ATP when 5-C acid is converted to 4-C acid.)
- In the next step, 4-C Succinic acid is oxidized (by removal of hydrogen) to 4-C fumaric acid in the presence of co-enzyme FAD (flavin adenine dinucleotide). A reduced $FADH_2$ is formed.
- Fumaric acid (4-C) is converted to another 4-C acid, malic acid, by the addition of H_2O.
- In the final step of the TCA cycle, 4-C Malic acid is oxidized (by removal of hydrogen) to 4-C Oxaloacetic acid. $NADH_2$ is formed in the process.

Thus, OAA is regenerated in the last step. It can combine withanother 2-c Acetyl Co-A to form Citric acid and participate in the TCA cycle again. During various steps of the Kreb's cycle, oxidation of substrate takes place by the removal of an electron hydrogen ion from substrate. It is accepted at each step by a suitable co-enzyme such as NAD or FAD to form a reduced co-enzyme molecule, $NADH_2$ or $FADH_2$ respectively.

The chart of the Kreb's cycle indicates the result of the participation of only one molecule of pyruvic acid. It should be noted that 2 pyruvic acid molecules are formed at the end of glycolysis. Therefore, the final analysis of the Kreb's cycle involving 2 pyruvic acids may be summarized as follows.

- Total number of reduced co-enzyme molecules formed = 8NADH2 and 2FADH2
- Total number of ATP formed directly (when 5 - C acid is converted to 4 - C acid) = 2ATP
- Total number of CO_2 molecules released = $6CO_2$
- Total number of H_2O molecules utilized = $6H_2O$

ELECTRON TRANSPORT CHAIN

This is the final part of the phase-II of aerobic respiration. As stated earlier, in respiration, oxidation of the substrate occurs by dehydrogenation (i.e., removal) of hydrogen atoms (2H) from the substrate. These are mostly accepted by NAD to form reduced co-enzyme NADH. In the aerobic respiration of a molecule of glucose (6-C), a total of, 10 $NADH_2$ are formed, (2$NADH_2$ in glycolysis + 8 $NADH_2$ in Kreb's cycle).

At one step in Kreb's cycle, hydrogen is accepted by FAD to form $FADH_2$. total, 2$FADH_2$ are formed in the aerobic respiration of each glucose. Each molecule of reduced co-enzyme thus formed in aerobic respiration (glycolysis and Kreb's cycle) is finally oxidized by the free molecular oxygen through a process called terminal oxidation.

$$\left.\begin{array}{l} NADH_2 + \frac{1}{2}O_2 \longrightarrow H_2O + NAD \\ FADH_2 + \frac{1}{2}O_2 \longrightarrow H_2O + FAD \end{array}\right\} \text{Terminal oxidation}$$

This involves participation of the respiratory chain, also called electron transport system (ETS).

The respiratory chain (or the ETS) is present in the inner membrane of mitochondrion (i.e., in the cristae membrane). It consists of various enzymes and co-enzymes which act as electron carriers. The hydrogen ions (and the electrons of hydrogen) from the oxidized substrate pass through the ETS and finally react with free molecular oxygen to form water. The electron flow from the substrate(NADH2 or FADH2)to O_2

(through ETS) is the source of energy for the formation of ATP. This is called oxidative phosphorylation.

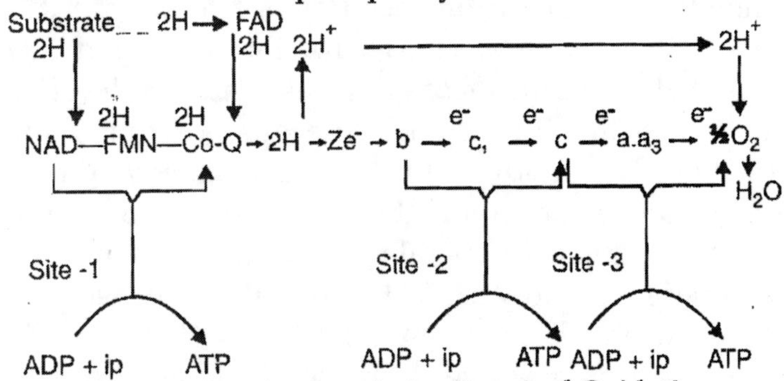

Fig. The Respiratory chain, Terminal Oxidation and oxidative phosphorylation.

Transfer of hydrogen and electrons through ETS

As stated earlier, the most common acceptor in respiration reactions is NAD. Thus, stet is transferred to Co-Q via-FMN (Flavin mononucleotide). At one step in Kreb's cycle, hydrogen is accepted by FAD. From FAD, it is transferred to Co-Q.

At Co-Q, the hydrogen undergoes ionization, i.e. it splits into electrons and protons.

$2H \rightarrow 2e^- + 2H^+$

Therefore, from Co-Q onwards, only the electrons are transferred through the ETS along the chain of the electron carrying molecules. The protons (H^+) are transferred across the inner membrane into the intermembrane space. The first electron acceptor in the chain is cytochrome. b. It accepts the electrons from Co-Q and transfers them first to cytochrome C1, and then to cytochrome C1. From cytochrom C1, electrons are accepted by cytochrome oxidase. In the final step, cytochrome oxidase donates electrons to the free molecular oxygen. In other words, oxygen acts as the last electron acceptor in the respiratory chain (aerobic respiration). Oxygen receives 2 electrons through the ETS and 2 protons directly from the aqueous medium to form one molecule of water.

$1/2O_2 + 2e^- + 2H^+ \rightarrow H_2O$ (Terminal oxidation)

Utilization of oxygen at the end of the ETS is the final step in aerobic respiration and is called terminal oxidation.

OXIDATIVE PHOSHORYLATION

The electron flow from the substrate to oxygen through the ETS is the source of energy for the formation of ATP. Transfer of electrons by these enzymes results in the HT being pumped into the intermembrane space from oxidative phosphation. The buildup of HT in the intermembrane space yield a concentration gradient, such that HT flow back into the stroma through the membrane-bound protien enzymes, ATP synthase. The flow of HT through the synthase provides energy for the synthase to convert ADP to ATP.

There is a progressive decrease in the level of electron energy as they flow along the Chain. This is given out as free energy. There are three sites in the ETS at which enough free energy is released. At each site, this energy is utilized to form ATP from ADP and iP (i.e. inorganic phosphate = H_3PO_4). Formation of ATP using the energy released during terminal oxidation is called oxidative phosphorylation. The respiratory chain, each molecule of reduced co-enzyme $NADH_2$ produces 3ATP through the ETS, while each FADH2 yields 2ATP through the ETS.

Table : Total ATP formed in aerobic respiration.

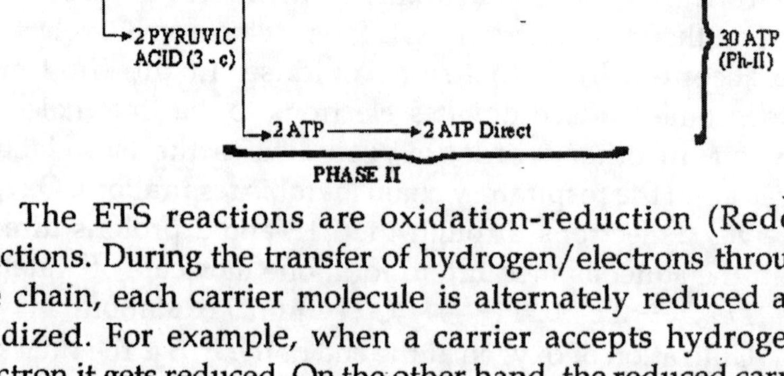

The ETS reactions are oxidation-reduction (Redox) reactions. During the transfer of hydrogen/electrons through the chain, each carrier molecule is alternately reduced and oxidized. For example, when a carrier accepts hydrogen/ electron it gets reduced. On the other hand, the reduced carrier is oxidized when it transfers the hydrogen/electron to the next carrier in the chain. There is a net flow of electrons through the ETS from more electronegative redox potential (i.e. from $NADH_2$ or $FADH_2$) to the more electropositive redox potential (i.e. to oxygen).

Overall Result of Aerobic Respiration

Complete oxidation of one molecule of 6-C glucose through aerobic respiration (involving glycolysis, Kreb's cycle, ETS, etc.) results in –

- Release of $6CO_2$
- Utilization of $6O_2$ and
- Formation of $6H_2O$.

$$\left.\begin{array}{l} 10\ NADH_2 + 5O_2 \longrightarrow 10H_2O + 10\ NAD \\ 2\ FADH_2 + O_2 \longrightarrow 2H_2O + 2\ FAD \end{array}\right\} \text{Through Terminal oxidation}$$

Total used $6O_2$ — $12\ H_2O$ formed.

However, $6H_2O$ are used up in Kreb's cycle (involving 2 pyruvic acid molecules).

$12\ H_2O$ formed - $6H_2O$ used = $6H_2O$ net gain

ANAEROBIC RESPIRATION

An intracellular-oxidation of organic substrate (e.g. glucose) without utilization of free molecular oxygen (in the absence of oxygen). The oxidation is incomplete and results in the formation of either ethyl alcohol or lactic acid, carbon dioxide and energy (2ATP).

Important Features

- Complete process of anaerobic respiration takes place in the cell cytoplasm only.
- Mitochondria are not involved.
- Oxygen is not used.
- Kreb's cycle reactions, ETS or oxidative phosphorylation are not involved.
- Phase-I consists of the same reactions of glycolysis (EMP pathway). found in aerobic respiration.
- Phase-II consists of decarboxylation and reduction reactions converting the pyruvic acid into either ethyl alcohol with the evolution of Carbon Dioxide.

The whole process may be summarized as follows:

$$(1)\ \underset{\text{glucose}}{C_6H_{12}O_6} \xrightarrow{\text{Glycolysis}} \underset{\text{pyruvic acid}}{2\,CH_3CO.COOH} + 2\,NADH_2 + 2\,ATP$$

Further anaerobic breakdown of pyruvic acid involves following two steps.

$$(2)\quad \underset{\text{pyruvic acid}}{2\,CH_3COCOOH} \xrightarrow{\text{Decarboxylation}} \underset{\text{acetaldehyde}}{2\,CH_3CHO} + 2\,CO_2$$

$$(3)\quad \underset{\text{acetaldehyde}}{2\,CH_3CHO} + 2\,NADH_2 \xrightarrow{\text{Reduction}} \underset{\text{ethyl alcohol}}{2\,CH_3CH_2OH} + 2NAD$$

After adding up the equations (1) + (2) + (3), the final overall equation for the anaerobic respiration is obtained as follows:

$$C_6H_{12}O_6 \longrightarrow \underset{\text{ethyl alcohol}}{2\,CH_3CH_2OH} + 2\,CO_2 + 2ATP$$

The 2 NADH2 gained in glycolysis are used in the reduction of acetaldehyde to ethanol.

FERMENTATION

Fermentation is of different types and takes place under anaerobic conditions mostly in saprophytic microorganisms like certain bacteria and fungi. However, it may also take place in higher organisms under certain conditions.

The two most common types of fermentation are:

- Alcoholic fermentation
- Lactic acid fermentation.

Alcoholic Fermentation

The type of fermentation in which ethyl alcohol is the main end product.This is very common in yeast (unicellular fungus) and also seen in some bacteria. Yeast cells release enzymes called zymase complex which bring about the fermentation. The reactions are similar to anaerobic respiration.

$$C_6H_{12}O_6 \xrightarrow[\text{Zymase}]{\text{YEAST}} \underset{\text{ethyl alcohol}}{2C_2H_5OH} + 2CO_2 + 2ATP$$

Lactic acid Fermentation

The type of fermentation in which lactic acid is the end product.

It is carried out by some bacteria (e.g. lactic acid bacteria), and also by animals (muscle glycolysis in animals, under oxygen deficiency, results in the formation of lactic acid this is whay we experience in muscle cramps, or "Charley horse").

Lactic acid bacteria can ferment milk sugar lactose ($C_{12}H_{22}O_{11}$) to lactic acid. The process is extracellular.

1. $$\underset{lactose}{C_{12}H_{22}O_{11}+11+H_2O} \xrightarrow[bacteria]{Lacticacid} \underset{hexose\ sugars}{C_6H_{12}O6+C_6H_{12}O_6}$$

2. $$C_6H_{12}O_6 \xrightarrow{Glycolysis} \underset{pyruvic\ acid}{2CH_3COCOOH}+2NADH_2+2ATP$$

3. $$2CH_3COCOOH+2NADH_2 \rightarrow \underset{lactic\ acid}{2CH_3CHOH.COOH}+2NAD$$

This brings about curdling of milk.

Commercial Application of Fermentation

There are various kinds of fermentation carried on by different microorganisms. Many of these result in highly useful end products. Such useful microbial activity is used on large industrial scale to obtain the useful end products for the benefit of mankind. Some of the industrial products of the microbial fermentation activities are:

- Antibiotics
- Vitamins
- Industrial alcohol
- Bakery products
- Some dairy products
- Tanning of leather
- Curing of tea and coffee
- Lactic acid
- Butric acid,
- Acetic acid etc.

SIGNIFICANCE OF RESPIRATION

- It is the energy releasing and supplying process in all living organisms.
- It converts food energy into metabolically usable forms of chemical energy.
- Respiration releases energy in a controlled and step-wise manner as a result of which most of the released energy is properly utilized for the cellular activities (e.g. metabolism, cell division, growth, etc.) Only a small part of energy is lost as heat.
- In other words, respiration conserves energy very efficiently. Out of the total 673 Kcal energy released per molecule of glucose, 456 Kcal is conserved as 38 ATP molecules.
- Various intermediate products of glycolysis and Kreb's cycle are used for biosynthesis of other complex organic compounds in cell metabolism.
- Carbon dioxide, required for photosynthesis by green plants, is replenished by carbon dioxide released in respiration. Similarly, oxygen required for respiration is replenished by oxygen released in photosynthesis. Thus, photosynthesis and respiration are complementary to each other and together help to maintain the balance of oxygen and carbon dioxide in nature.
- Fermentation, which is a type of anaerobic respiration, is helpful in the industrial production of various useful products such as alcohol, antibiotics, vitamins, organic acids, bakery and dairy products, tanned leather, etc.

Chapter 3

Genes and Molecular Genetics

Mendel was the first to put forward clearly the idea of hereditary units, and referred to them as 'factors'. These were later named genes. In 1903, Sutton and Bovery postulated the chromosomal theory of inheritance and stated that the genes were located in the chromosomes. This was subsequently confirmed by the works of Morgan and others and is now accepted universally. Thus, it was established that the chromosomes are the carriers of the genetic material. The chemical of which the genes are composed is called the genetic material.

QUEST FOR THE IDENTIFICATION OF THE GENETIC MATERIAL

The nucleic acids were discovered by Meischer in 1869. In later years, it was also established that chromosomes are made up of proteins (histones and protamines) and nucleic acids (DNA). Now, the question was, which of these actually represented the genetic material? Was it the proteins or the DNA? The process of identification of genetic material began in 1928 and was successfully concluded in 1952. It was conclusively established that it is the DNA that contains the genetic material and not the proteins in the chromosomes. This fact could be established only through some brilliantly conceived and designed novel experiments. The most notable among these are:

- *Experiments of Griffith (1928):* Through his experiments, Griffith discovered the phenomenon of transformation (a special type of genetic recombination) in which the non-virulent strain of the bacterium Diplococcus pneumoniae was transformed into the virulent strain. Griffith's experiments started the process of the identification of the genetic material.

- *Avery, Macleod and Mc Carty (1944):* repeated, elobourated and modified the transformation experiments of Griffith.

  They provided the first clear evidence the DNA is the genetic material and is responsible for the transformation in bacteria. They also gave evidence of the chemical nature of genes.
- *Lederberg and Zinder (1952):* confirmed the identity of DNA as the genetic material through a bacteriophage (vector), and the subsequent incorporation of the segment in the recipient DNA thereby causing transformation.
- *Hershey and Chase (1952):* confirmed that at least in the bacterial virus called T2-phage, DNA is the only genetic material. This fact has been universally accepted since then.
- *Frankel - Conrai and Singer (1957):* established that in some viruses, RNA functions as the genetic material (and not the DNA).

Table: Some Additional Historical Events

Year	Scientists	Discovery/Event
1909	Johannsen	Coined the term "gene"
1924	Feulgen	Showed that chromosomes
	contain DNA	
1953	Waston and Crick	Double stranded helical
	model for structure of DNA	
molecule		

As stated earlier, chromosomes are the carriers of the hereditary material. In cells, chromosomes are located in the nucleus. In prokaryotic cells (e.g., bacteria, cyanobacteria, etc.), organized nuclei are not present. As a result, the hereditary material occurs in the cell cytoplasm as a nucleoid without the nuclear envelope.

PACKAGING OF HEREDITARY MATERIAL

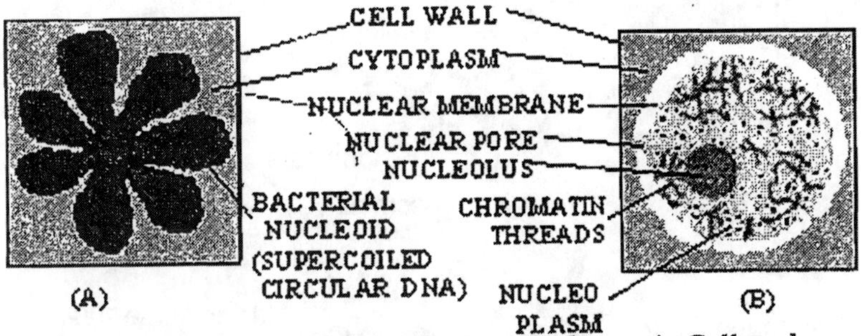

Fig. Structure of Nucleus (A) In a Prokaryotic Cell and (B) in an Eukaryotic cell

EUKARYOTIC NUCLEUS

In eukaryotic cells (e.g., all higher plants and animals), a well organized nucleus is present. It is bounded by a double membrane nuclear envelope with trilaminar structure. There are pores in the envelope. Internally, the nucleus is filled with nucleoplasm (nuclear sap) that is acidophilic and clear. The interphasic nucleus (i.e. the nucleus which is not in the stage of division) usually shows the presence of one dark, spherical body called nucleolus. The nucleoplasm also contains nuclear reticulum or chromatin network. It consists of very fine, long chromatin fibers. These represent the chromosomes at the interphasic stage. They become short, thick and distinct during cell division.

STRUCTURE OF CHROMATIN

The chromosomes are composed of nucleoprotein called chromatin. The chromatin fiber appears to have a structure like a 'string of beads.' According to the nucleosome-solenoid model proposed by Komberg and Thomas (1974), the beaded string is made of repeating units called nucleosomes. Each nucleosome is a bead-like particle about 60 Ao high and 110 Ao wide. In the nuclesome, the DNA double helix is wound around the core of eight histone molecules (octamer). The segments of DNA joining the 'beads' are called linker DNA. Each linker DNA has an average of one molecule of a histone protein (H1) attached to it

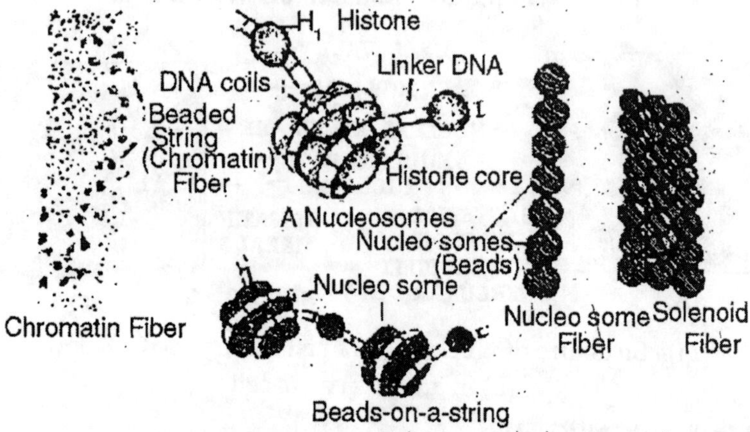

Fig. Nucleosome - Solenoid model for the Structure of Chromatin

The thin, long and "beads-on-string" chromatin fiber is condensed and packed into a short and thick chromosome (metaphase chromosome) because of further coiling and "super coiling", wherein (1) the string of beads condenses and coils to produce a 100 Ao thick nucleosome fiber (2) the nucleosome fibre then supercoils to produce a 300-350 Ao thick solenoid fiber. A solenoid represents the structure of chromatin fiber in the metaphase chromosome.

THE STRUCTURE OF DNA

Deoxyribonucleic nucleic Acid (DNA) is a highly complex megabiomolecule. The long chain molecule is formed of repeating units called nucleotides. Hence, it is described as the polynucleotide molecule. It consists of two polynucleotide antiparallel strands which are spirally coiled round each other along their lengths.

Chemical Components of DNA

The highly complex DNA molecule is composed of only three types of chemical components. These are (i) deoxyribose sugar, (ii) a phosphate, and (iii) nitrogen containing organic bases.

- Deoxyribose sugar: The sugar present in the DNA molecule is called deoxyribose sugar and hence the nucleic acid is called deoxyribonucleic acid (DNA). It is a pentose sugar (with 5 carbon atoms) having a pentagonal ring structure.

- Phosphate: The phosphate in the DNA is present as phosphoric acid (H_3PO_4). It has three reactive (-OH) groups of which two are involved in the formation of the sugar-phosphate backbone of each DNA strand.
- The nitrogen-containing organic bases

N-Bases
- Purines
  - Adenine (A)
  - Guanine (G)
- Pyrimidines
  - Cytosine (C)
  - Thymine (T)

These are heterocyclic compounds containing nitrogen in their rings and therefore called nitrogenous bases. DNA contains four different bases called adenine (A), guanine (G) cytosine (C), and thymine (T). These are grouped into two classes on the basis of their chemical structure: (i) Purines (with a double ring structure) and (ii) Pyrimidines (with a single ring structure).

Nucleosides

In the molecules of nucleic acids (DNA and RNA), each pentose sugar molecule has one nitrogen base attached at carbon number 1. It may be either a purine or a pyrimidine base. Thus, a pentose sugar with the attached N-base forms a nucleoside.

[Sugar + N-base] = Nucleoside.

In DNA, the deoxyribose sugar has one of the four bases (A,G,T or C) attached. Therefore, the nucleosides in DNA are called deoxyribosides.

[deoxyribose sugar + N-base] = Deoxyriboside

Nucleotides: A nucleoside with a phosphate group attached to it is called a nucleotide.

[Nucleoside + Phosphate] = Nucleotide

A nucleotide is the basic unit or monomer in the structure of a nucleic acid molecule. Thus, a nucleotide is a nucleoside phosphate. In a nucleotide, the phosphate group is linked with the pentose sugar at carbon-5. In the DNA, each nucleotide is called deoxyribonucleotide.

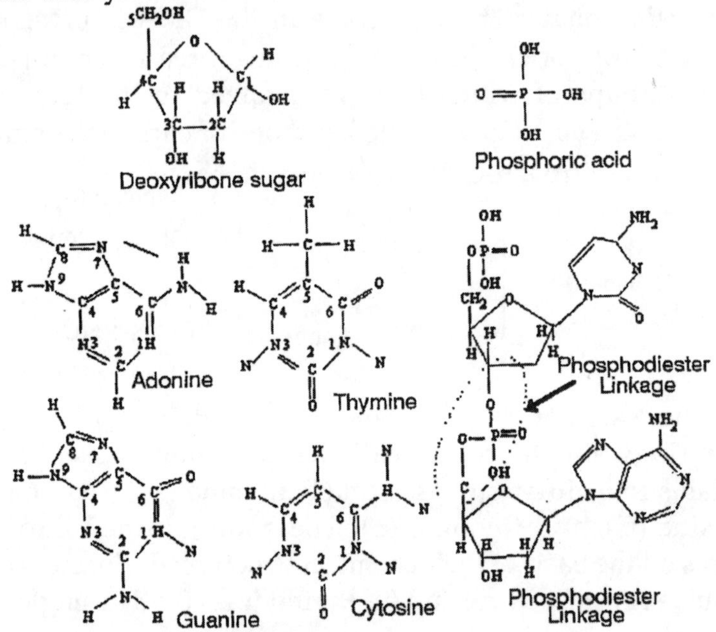

Fig. Chemical Components of DNA and the Phosphodiester Linkage

[Deoxyribose sugar + N-base + Phosphate] = Deoxyribonucleotide

As there are four different bases (A,T,G and C) in DNA, there can be only four types of nucleotides in DNA. The nucleotides act as the building block molecules for the synthesis of the polynucleotide molecules. When two nucleotides are linked together, a dinucleotide (or dimer) results. Similarly, when many nucleotides are linked with each other in a linear fashion, the resulting chain, is referred to as a polynucleotide (polymer, or). For example, each strand of DNA molecule has many deoxyribonucleotides linked in a chain-like arrangement. Therefore, it is described as polynucleotide strand and DNA molecule as polynucleotide molecule. In the DNA polynucleotide strands, the nucleotides are joined with each other by 'phosphodiester linkages'. In a phosphodiester linkage, the phosphate group present at the

C-5 of the sugar of one nucleotide gets attached to the C-3 of the sugar of the next nucleotide in the chain.

The Watson-Crick model of the DNA double helix

The most widely accepted model for the structure of DNA molecule was proposed by Watson and Crick in 1953. According to their model, the DNA has the following structural characteristics.

Molecule

The DNA molecule is a double helix. The molecule is formed by two antiparallel polynucleotide strands which are spirally coiled round each other in a right-handed helix. The two strands are held together by hydrogen bonds. The double stranded helical molecule has alternate major (or deep) and minor grooves.

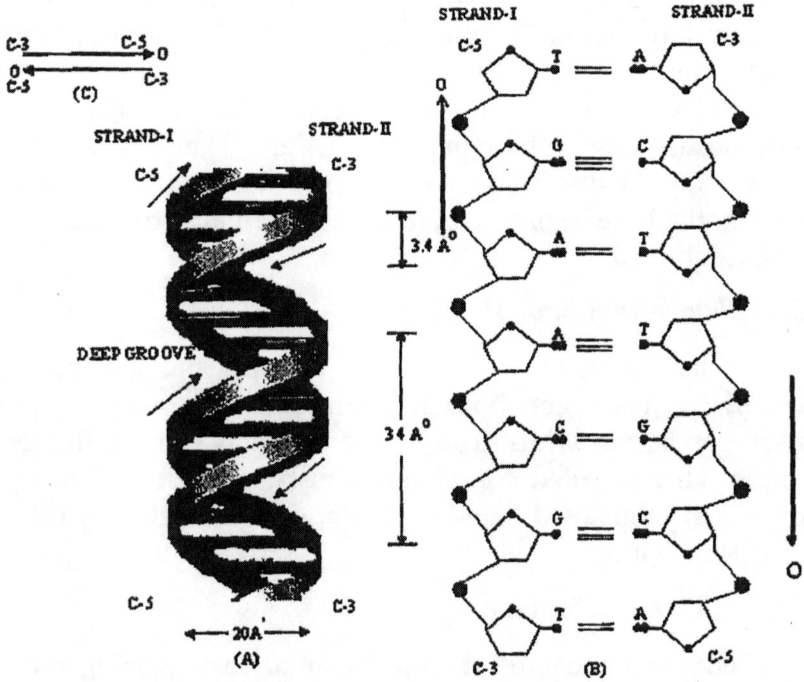

Fig. Structure of DNA (Watson and Crick model). (A) DNA double helix. (B) Detailed Structure of the two Strands. (C) C-5 and C-3 ends and Antiparallel Nature of Strands (Diagrammatic)

Structure of Each Strand

Each strand is a long polynucleotide of deoxyribonucleotides. The backbone of the strand is formed by alternately arranged deoxyribose sugar and phosphate molecules which are joined by the phosphodiester linkages. Each sugar in the strand has one base horizontally attached to it at carbon-1. It can be any one of the four: A,T,G or C. These four N-bases can occur in any possible sequence along the length of a strand.

Complementary Nature of the Strands

The two strands are complementary to each other with regards to the arrangement of the bases in the two strands. For example, where adenine (a purine) occurs in one strand, thymine (a pyrimidine) is present in the corresponding position in the opposite strand and vice versa.

Similarly, wherever guanine (a purine) is present in one strand, the other strand has cytosine (a pyrimidine) opposite to it and vice versa. Thus, in the double helix, purines and pyrimidines exist in base pairs, i.e., (A and T) and (G and C). As a result, if the base sequence of one strand of DNA is known, the base sequence of its complementary strand can be easily deduced.

Complementary Base Pairing

In each pair, the two bases of the opposite strands are joined by hydrogen bonds. A and T are joined by two hydrogen bonds, while G and C are joined by three hydrogen bonds. This is called complementary base pairing. The two strands are thus held together all along their lengths by these hydrogen bonds.

Purine: Pyrimidine Ratio

Because of the fixed or complementary base pairing in the DNA molecule, the total number of A is equal to the total number of T and the total number of G is equal to the total number of C. In other words, (A+G)= (T+C). Hence, purines: pyrimidines ratio is 1:1.

C-3 and C-5 Ends of the Strand

In each strand one end of the strand has one free phosphate group on carbon-5 of the sugar molecule. This is the end of the strand is called C-5 (or 5') end. The other end of the strand has a free -OH on carbon-3 of the sugar molecule. This is called C-3 (or 3') end of the strand.

Antiparallel Nature of Strands

The two strands are oppositely oriented and hence are called antiparallel. This means, the 3'end of one strand is adjacent to the 5'end of the other strand. This is because, the phosphate-sugar linkages run in opposite directions in the two strands.

Dimensions

The diameter of the DNA double helix is 20 A°. The length of each complete spiral (turn or pitch) of the molecule measures 34 A°. 10 pairs of nucleotides are present in each complete spiral. Therefore, each nucleotide in the strand occupies a distance of 3.4A^0.

Circular DNA Molecules

Chromosomes of most prokaryotes (e.g. bacteria, cyanobacteria,) are circular molecules of DNA. In bacteria, there is no organized nucleus. The bacterial nucleoid consists of a single circular DNA molecule (bacterial chromosome). The molecule has two complementary strands forming a covalently closed circle. Generally, the circular molecule is present in a highly folded and suspercoiled state. This is expected because the diameter of a bacterial cell (e.g. Escherichia coli) is about 1-2 microns while the total length of the circular DNA is about 1100 microns. The circular molecule has 40-50 folds or looped domains. These folds are held in position by RNA molecules (RNA connectors) and some non-histone proteins associated with the bacterial chromosome. (Histones are absent in bacteria).

The DNA segment in each loop is supercoiled independently. Because of this characteristic formation of loops as well as the supercoils within the loops, the large circular DNA molecule can be packed into a small bacterial cell. Otherwise, the relaxed and fully expanded circular molecule would be too large (about 350 microns diameter) to be contained in the bacterial cell. The coils of the supercoiled circle can be relaxed by treatment with enzymes such as RNAse or DNAse. The uncoiling occurs due to a break (nick) in one or both the strands of DNA. In some viruses, e.g. certain bacteriophages, the circular DNA is single stranded. It becomes double stranded only during replication (replicatable form).

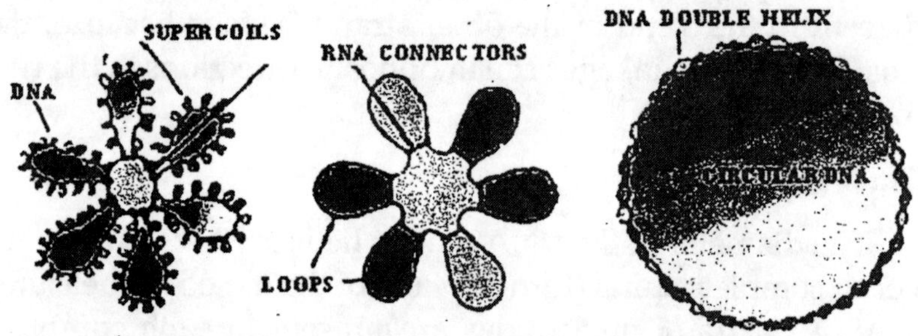

Fig. (A) Supercoiled Folded Circular DNA, (B) Folded Circular Molecule, (c) Fully Relaxed Circular DNA Molecule

REPLICATION OF DNA IN EUKARYOTES

Definition: "The process by which DNA produces daughter DNA molecules which are exact copies of the original DNA is called replication of DNA." In eukaryotes, DNA is double stranded. The two strands are complementary to each other because of their base sequences.

Semi-conservative Method of DNA Replication

Important points:

- This is the most common method of DNA replication.
- It takes place in the nucleus where the DNA is present in the chromosomes.
- Replication takes place in the S-phase (synthesis phase) of the interphase nucleus.
- The deoxyribose nucleotides needed for the formation of the new DNA strands are present in the nucleoplasm.

At the time of replication, the two strands of DNA first separate. Each strand then acts as a template for the formation of a new strand. A new strand is constructed on each old strand, and two exactly identical double stranded DNA molecules are formed. In each new DNA molecule, one strand is old (original) while the other is newly formed. Hence, Watson and Crick described this method as semi-conservative replication.

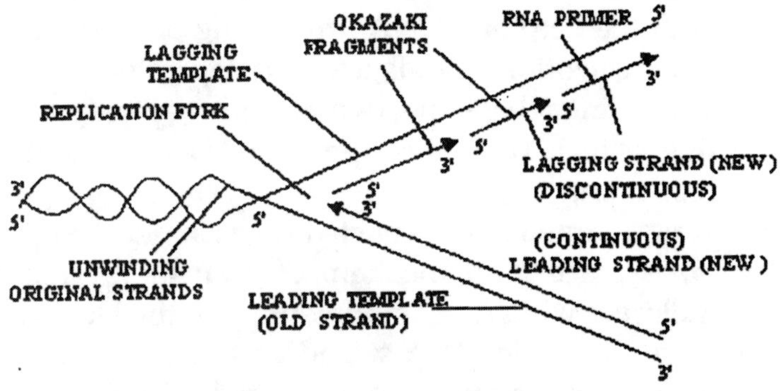

Fig. An overall Process of DNA Replication Showing Replication fork and Formation of new Strands Template and Lagging Template

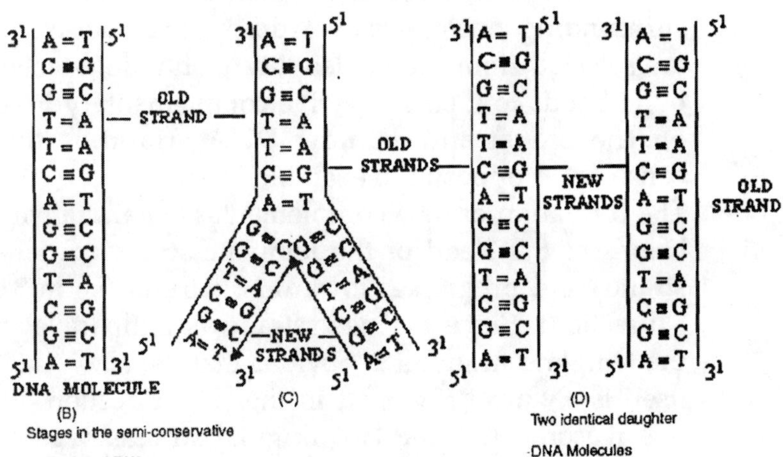

Fig. Stages in the Semi-Conservative Method of DNA Replication in Eukaryotes

The various steps involved in this process are summarized as follows:

- The mechanism of replication starts at a specific point of the DNA molecule. This is called the origin.
- At the origin, the DNA strand breaks because of an incision (nick). This is made by an enzyme called incision enzyme (endonuclease).
- The hydrogen bonds joining the two strands are broken by the enzyme.
- The two strands start unwinding. This takes place with the help of a DNA unwinding protein. The two polynucleotide strands are thus separated.
- The point where the two strands separate appears like a fork or a y-shape. This is described as a replicating fork.
- A new strand is constructed on each old strand. This takes place with the help of a small RNA primer molecule which is complimentary to the DNA at that point. Each old DNA strand acts as a template (site) for the construction of new strand. The RNA primer attaches itself to the old strand and attracts the enzymes which add new nucleotides. The deoxyribose nucleotides are present in the surrounding nucleoplasm. Appropriate nucleotides are selected from the nucleoplasm, and are attached by H-bonds to their respective complementary bases on the old strand. A new DNA strand is thus constructed opposite to each old strand.
- The formation of new complementary strand always begins at the 3'end of the template strand (original strand) and progresses towards the 5'end (i.e in 3'® 5'direction). Since the new strand is antiparallel to the template strand, it is obvious that the new strand itself is always developed in the, 5'®3'direction. For this reason when the two original strands separate (then with respect to the origin of separation), one acts as 3'®5'template while the other acts as 3'®5'template. Of the two, the replication of 3'®5'template begins first. Hence the new strand formed on it is called the leading strand. The other

template (5'®3') must begin replication at the fork and progress back toward the previously transcribed fragment. The new strand formed on it is called the lagging strand.

- Replication of the lagging strand takes place in small fragments called Okazaki fragments. These are then connected together by the enzyme polynucleotide ligase.
- Replication may take place in only one direction on the DNA helix (unidirectional) or in two directions (bidirectional).
- At the end of the process, two double stranded DNA molecules are formed from the original DNA molecule.
- In each newly formed DNA, one strand is old while the other is new. Hence, it is described as semi-conservative replication.

The semi-conservative nature of DNA replication was confirmed by Meselson and Stahl with the help of an experiment. They marked the DNA in Esherichia coli with heavy isotope of nitrogen (15N) and then traced it in the following generations of the E-coli progeny. It was demonstrated that every daughter DNA is an exact copy of the parent DNA and that each daughter DNA has one strand of the parent (old strand) and the other strand in newly formed.

REPLICATION OF PROKARYOTIC CHROMOSOME

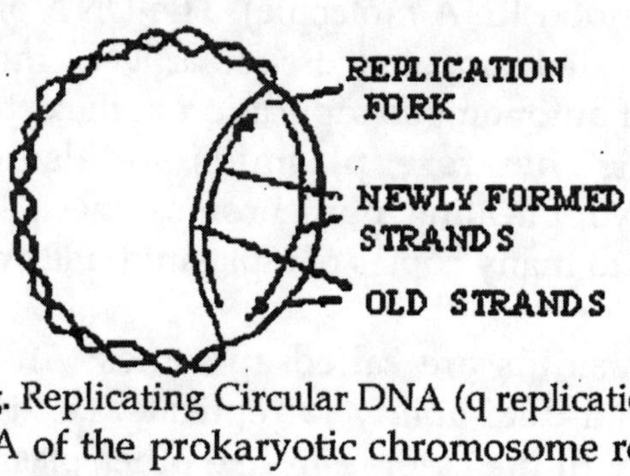

Fig. Replicating Circular DNA (q replication)

The DNA of the prokaryotic chromosome replicates as a circular structure. It is a semiconservative as in the linear DNA of the eukaryotes. Replication begins a fixed point called the origin. In most prokaryotes, the DNA has only one 'origin' point. The uncoiling of the two strands of the circular molecule begins at the point of origin and progresses in opposite directions (bidirectional). This is helped by the enzyme called DNA gyrase.

Simultaneously with the uncoiling of the original strands, a new complementary strand is being constructed on each strand. As a result, the circular DNA appears as the Greek letter q (theta) during replication. Hence this mode of replication is called q replication.

Each newly forming strand also goes on helically coiling around its circular template strand (original strand). As a consequence, by the time the replication process is over, two complete double-stranded circular DNA molecules are formed. These are exact copies of the original molecule. Moreover each daughter molecule has one original (parental) and the other newly formed circular strands.

Plasmids

In addition to the normal DNA chromosome, extrachromosomal genetic elements are often found in the cytoplasm of most bacterial species and in some species of eukaryotes.

These elements are called plasmids. Plasmids are extrachromosomal circular DNA molecules (except the plasmid called Killer-Plasmid present in yeast cells, which is a double stranded RNA molecule). The DNA of all plasmids is double stranded supercoiled circular DNA molecule. These are capable of autonomous replication in the cytoplasm of the bacterial cell. Therefore, plasmids are also described as autonomously replicating 'minichromosomes'. A bacterial cell can have one to many copies of a plasmid. Plasmids represent extra genes.

Some plasmids are called episomes. An episome is a plasmid with a dual ability to replicate, i.e. it can replicate autonomously in the cytoplasm like other plasmids or it can also integrate itself into the DNA of the bacterial chromosome and behave as part of the chromosome.

Role of Plasmids

Many plasmids contain important and essential plasmid genes which are helpful under certain conditions. For example, a bacterial cell possessing R Plasmid develops resistance to certain antibiotics. Similarly, presence of some plasmids gives the bacterial cell the ability to produce toxins (toxigenicity), or the ability to use unusual chemical compounds as nutrients. Moreover, plasmids and episomes prove to be good vehicles for gene manipulations (genetic engineering).

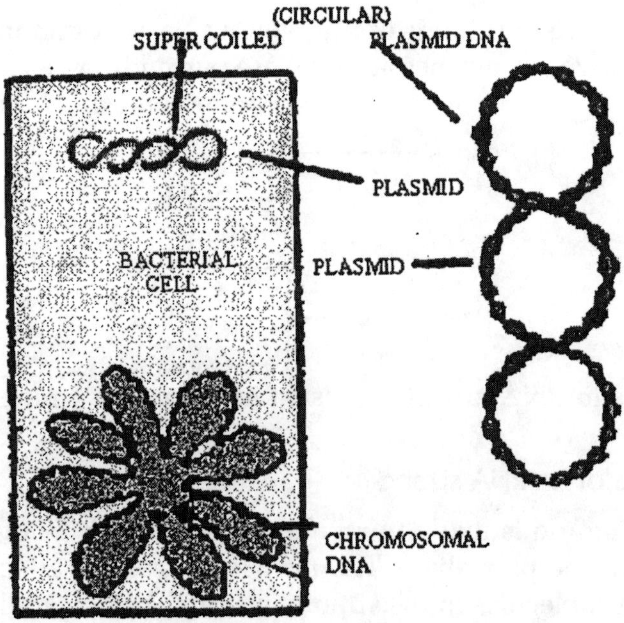

Fig. Bacterial Cell Showing a Plasmid

RNA: STRUCTURE AND TYPES

Ribonucleic Acid (RNA) is another polynucleotide which occurs in the cells as non-genetic material, with the exception of some viruses. RNA is present in the nucleus as well as in the cytoplasm.

General Structure

RNA molecule is single stranded and consists of

nucleotides arranged in a long series. The single strand of RNA may be simple and straight, or it may be variously folded upon itself in certain regions.

Structural components

RNA molecule has three primary components.

- Ribose sugar (a pentose sugar), with a pentagonal ring structure
- Phosphate, as phosphoric acid
- Nitrogenous bases

There are four kinds of nitrogenous bases found in RNA. Of those, two are purines and two are pyrimidines, as follows:

(I) Purines
- Adenine (A)
- Guanine (G)

(II) Pyrimidines
- Uracil (U)
- Cytosine (C)

Thus in RNA, uracil is present in place of thymine found in DNA.

Structure of an RNA strand

The strand is made up of alternating molecules of ribose sugar and the phosphate. The nitrogen bases are attached to the sugar molecules in the strand and 'stick out' laterally as in DNA. A sugar, a N-base and a phosphate together form a ribonucleotide.

A nucleotide without the phosphate is called ribonucleoside. RNA being single stranded, the nitrogen bases remain mostly unpaired. However, the strand may be folded upon itself in certain regions. In such folded regions, base pairing occurs between purines and pyrimidines as follows:

Adenine = Uracil (two H-bonds). Guanine = Cytosine (three H-bonds).

Nitrogen bases remain unpaired in the unfolded regions of the strand. Because of this variability in base pairing in different regions of the same strand, the total number of purines need not be equal to the total number of pyrimidines in RNA.

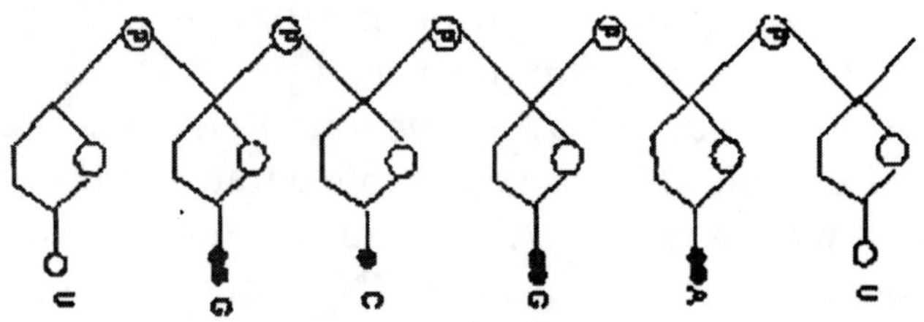

Fig. General Structure of RNA

DIFFERENCE BETWEEN DNA AND RNA

	Characters	DNA	RNA
1.	Molecule	Double stranded, helical	Single stranded, straight or variously folded and twisted.
2.	Pentose sugar	Deoxyribose	Ribose
3.	Pyrimidine base	Thymine	Uracil
4.	Complementary base pairing	Always present and exists between A = T and G = C	Normally absent, but may be present in twisted segments of a molecule. If present, pairing is between A = U and G = C
5.	Ratio of Purines: Pyrimidines	Always 1 : 1	Not necessarily 1:1
6.	Replication	Can replicate	Cannot replicate

TYPES OF NON-GENETIC RNA AND THEIR FUNCTIONS

There are three types of non-genetic RNA.

- mRNA or messenger RNA,
- rRNA or ribosomal RNA and
- tRNA or transfer RNA.

mRNA (Messenger RNA)

This is called messenger RNA because it carries information for protein synthesis from the DNA to the ribosomes in the cytoplasm (the site of protein synthesis). m-RNA constitutes about 3-5% of the total RNA. It is produced on the DNA strand. The process is called transcription. Hence, the base sequence of mRNA is complementary to that of the DNA strand. The bases on the mRNA strand are organized into triplets. Each triplet consists of a sequence of three consecutive bases (nucleotides) and is called a codon (code word). Each codon specifies one amino acid.

The sequence of codons on the mRNA strand is called the mRNA language. It indicates the sequence of amino acids for the synthesis of a protein. It begins with the codon AUG (initiation codon or starting codon) and ends with either UAA, UAG or UGA (stop codons). The single-stranded mRNA molecule is always straight and therefore, base pairing is totally absent in mRNA.

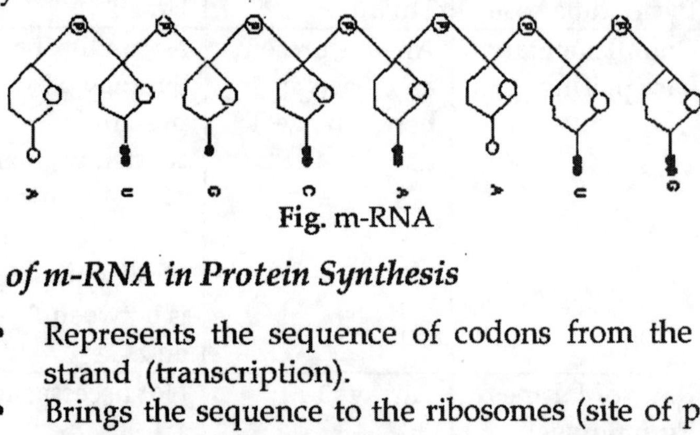

Fig. m-RNA

Role of m-RNA in Protein Synthesis

- Represents the sequence of codons from the DNA strand (transcription).
- Brings the sequence to the ribosomes (site of protein synthesis) in the cytoplasm.
- Provides the sequence for the synthesis of specific protein from the amino acids with the help of t-RNA (translation).

rRNA (Ribosomal RNA)

r RNA forms about 80% of the total RNA. It is present in the ribosomes in the cell cytoplasm (site of protein synthesis) and hence called rRNA. The single-stranded molecule of rRNA is variously folded and twisted upon itself in certain regions. In such folded regions, complementary bases form pairs and are joined by hydrogen bonds.

Role of rRNA in Protein Synthesis

The role of rRNA in protein synthesis is not yet very clearly known but it is known to complex with various protiens. The resulting structure is a ribosome, and this complex reads the coded sequence in mRNA to link amino acids together into particular protiens.

- It provides proper binding sites for the mRNA of the ribosomes.
- It orients the mRNA in such a way that its nitrogen base triplets or codons are properly read or translated.
- It also releases tRNA after the transfer of activated amino acid.
- It protects the mRNA strand from the action of enzymes (nucleases).
- It protects the growing (nascent) polypeptide chain from proteolytic enzymes.

tRNA (Transfer RNA)

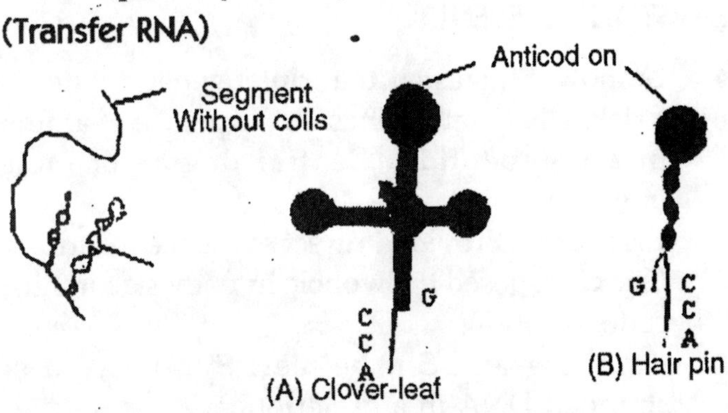

(A) Clover-leaf (B) Hair pin

It is the smallest of all the types of RNA. About 10 to 20% of the total RNA of the cell is of this type. tRNA strand is folded upon itself forming loops. It results in either a clover leaf pattern or hair pin pattern. One end of the strand has guanine, while the other end carries the CCA combination of nitrogen bases. A triplet of nitrogen bases called anticodon is present on one of the loops.

The anticodon pairs with the complementary codon on the mRNA molecule. The tRNA molecules carry amino acids to the mRNA during the process of protein synthesis. Each type of the amino acid is carried by a specific tRNA molecule. tRNA is synthesized on the DNA template. It has complementary base pairs in folded regions.

Role of tRNA in protein synthesis

- tRNA carries the required specific amino acids from cell cytoplasm to the ribosome (site of protein synthesis).
- Each type of amino acid is carried by a specific type of tRNA.
- In the ribosome, tRNA helps to arrange the amino acids in their proper sequence for the synthesis of a protein. This is done with the help of the codons on the mRNA and the matching (complementary) anticodons on the tRNA (translation).

THE GENETIC CODE

SOME HISTORICAL EVENTS

1954 Gamow: Suggested the triplet genetic code.

1958 Crick: DNA determines the sequence of amino acids in a polypeptide (Central dogma of molecular biology).

1961 Crick et al: Provided direct evidence for triplet code.

1966 Crick: Proposed the wobble hypothesis for the genetic code.

1967 Komberg and Sinsheimer: Synthesized single-stranded DNA in a bacteriophage.

1970 Khorana: Synthesized an artificial gene from DNA nucleotides.

THE GENETIC CODE

DNA is genetic material and contains genetic information. The expression of a gene takes place through specific enzymes. Each gene produces a specific (one-gene one-protien hypothesis). In other words, formation of each specific protein is controlled by a particular gene. A gene is (almost always) a segment of DNA strand and so, the information for the formation of a protein is contained in the DNA strand.

Further, each protein is a long polypeptide chain molecule formed by joining amino acid molecules. From the cell pool, only 20 different types of amino acids are used for protein synthesis. The sequence of the nitrogen bases in the DNA determines the sequence of amino acids in a protein molecule through the mRNA.

This sequence is copied down by the mRNA (transcription). It is present on the mRNA strand in the form of coded language (cyptogram or mRNA language or genetic code). The mRNA bases (A, U, C and G) serve as the four alphabets of the coded language.

Codon

The smallest sequence of the nitrogen bases (nucleotides) on the mRNA which can specify one amino acid is called a codon. Each codon consists of three successive bases on the mRNA. Why should each codon in the genetic code consist of 3 bases (triplet codon) and not of one base each or 2 bases each? This is because there are 20 different amino acids which can be used in the synthesis of proteins in the cells. There must be at least one specific codon for each amino acid. Thus, there has to be at least 20 different codons in the genetic code. There are only four bases. A minimum of 3 bases per codon is necessary to have (a minimum of) 20 codes.

The Wobble Hypothesis

The anticodon on tRNA is complementary to the codon on the mRNA as per the A = U, G = C base pairing rule. However, it has been observed that the 3rd base position may vary and yet still code for the same amino acid. For example, both codons TTA and TTG code for the amino acid, leucine. Thus, the third position is called the wobble position. Thus, Crick's wobble hypothesis explains the degeneracy of the genetic code at the third position of the codon.

THE CENTRAL THEME OF PROTEIN SYNTHESIS

A protein is a long chain polypeptide. The chain molecule is formed by joining amino-acid molecules with peptide linkages. The sequence of amino acids in each type of protein is highly specific. The functional proteins form enzymes and hormones and control all metabolic and biochemical reactions and processes in cells. Any alteration in the sequence of amino acids in protein molecules can affect the function of that protien. Hence, bio-synthesis of proteins is a highly specific process. The main principle of protein synthesis. DNA determines the sequence of amino acids in a polypeptide (protein) through mRNA. This is the main principle (central dogma) of protein synthesis. This involves transcription and translation.

$$\text{DNA(Strand)} \xrightarrow{\textit{Transcription}} \text{m} - \text{RNA} \xrightarrow{\textit{Translation}} \text{Protien}$$

Components involved in protein synthesis

The process requires.

- The 20 types of amino acids that are specified by the genetic code.
- DNA (to provide the base sequence).
- The non-genetic RNAs (m-RNA, t-RNA and r-RNA).
- Ribosomes (site of protein synthesis) and
- Various enzymes (factors).

Mechanism of protein synthesis

As stated earlier, the mechanism of protein synthesis involves two main events namely

- Transcription
- Translation.

Transcription

Transcription is the process of formation of mRNA on the DNA strand. It takes place in the presence of enzyme RNA Polymerase. One of the strands of DNA acts as a template for the formation of the same m-RNA in a series, one by one. Such ribosomes bound to the same mRNA strand are called polyribosomes or polysomes.

Translation

In Translation one particular group gets transmitted from one Band of DNA to another Band of DNA. In this procedure the DNA polymer (enzyme) is used. Normally, 40% of protein synthesis occurs in this manner.

Chapter 4

Mendellian Genetics

"Genetics is the study and understanding of the phenomena of heredity and variation." The term 'genetics' was first coined by Bateson in 1906. In Latin, it means genesis or origination of organisms. Heredity is the transmission of characters from one generation to the next, i.e., from parents to their offspring. Because of heredity, the offspring resemble their parents. Heredity is the essence of self-reproduction. It is owing to heredity or self-reproduction that we commonly observe the phenomenon of "like begets like", i.e., a seed of mango develops into a mango tree, or the offspring of a dog is a puppy, and that of human beings is a human being only.

Variations are the visible differences between the parents and the offspring, or between two offsprings of the same parents. An offspring receives all the characters from its parents and yet, an offspring is never an exact copy of its parents. Similarly, no two offsprings of the same parents are identical (exception: identical twins).

In order to understand the principles of inheritance and to discover the reasons for the variations, Mendel began a systematic search during the second half of the nineteenth century. For this, Mendel experimented on garden pea plants and performed various crosses with great precision, care and objectivity. He carefully counted the plants resulting from such crosses and kept statistical records of successive generations with the accuracy of a mathematician.

GREGOR MENDEL

Gregor Johann Mendel was born on July 22, 1822 in

Moravia, Austria. He had his early education in a monastery in Brunn, Austria (now Brno in Czechoslovakia) and later studied science and mathematics at the University of Vienna. He graduated in 1840. Mendel returned to the monastery in Brno as a monk. He worked as a teacher of physics and natural science in a Higher Secondary School of Brno during 1854 to 1868. He was appointed abbot of the monastery in 1868 and held this post untill his death.

Mendel carried out his legendary experiments on garden pea plants in the monastery garden from 1857 to 1865. He had a clear perspective and worked on the experiments with precision and thoroughness. He published his research paper containing his observations and conclusions in 1866 in the annual proceedings of the Natural History Society of Brunn. These conclusions are now known as Mendel's Laws. This work is a classic in biology for its elegance and simplicity and ranks amongst the most outstanding biological contributions of all times. But unfortunately, this work failed to attract the attention of the biologists of that time. Hence, it went ignored for 35 years. One of the possible reasons for such neglect was the inability of the biologists of that time to understand and appreciate the statistical approach adapted by Mendel. Thus, Mendel was left bitterly disappointed, and died an unrecognized death in 1884.

Sixteen years after Mendel's death, in 1900, Hugo de Vries (Holland), Karl Korrens (Germany) and Von Tschermark (Austria) independently arrived at similar conclusions as those of Mendel. De Vries rediscovered the research paper of Mendel and it was published again in 1901. The experiments on heredity of plants and animals since then have confirmed that Mendel's laws of heredity are applicable to other organisms as well. They form the basis of modern genetics. Hence, Mendel is called The Father of Genetics.

MENDEL'S EXPERIMENTS ON SWEET PEA

Selection of the Material

Mendel selected garden sweet pea (Pisum sativum) for his hybridization experiments for of the following reasons:

- Plants are annual and easy to cultivate.

- Peas have many distinct, well-defined and easily observable morphological characteristics (traits).
- Flowers are bisexual and naturally self-fertilizing, but they can also be easily cross-fertilized.
- The offspring of cross-fertilized plants are fertile.
- Flowers are sufficiently large for easy emasculation (removal of stamens) and artificial cross-pollination.

SELECTION OF CHARACTERS (TRAITS)

Mendel selected 14 different varieties of the pea and grouped them into seven pairs. Each pair was considered for a specific trait (characteristic) such as flower colour or seed shape or stem length, etc. The two members of each pair showed contrasting forms of the chosen trait,,e.g., in a pair selected for stem length, one variety had a tall stem (6-7 feet tall) while the other had a dwarf stem.

Table. Showing seven pairs of Contrasting Characters selected by Mendel

Sr. No.	Characters	Contrasting pairs (Allelic pairs)	
		(Dominant)	(RecBIOe)
1.	Form of seed	Round (R)	Wrinkled(r)
2.	Color of cotyledons	Yellow(Y)	Green(y)
3.	Color of seed coat	Colored(C)	White(c)
4.	Form of pod	Inflated (I)	Constricted(i)
5.	Color of pod	Green(G)	Yellow(g)
6.	Position of flower	Axial(A)	Terminal(a)
7.	Height of plant (Length of stem)	Tall(T)	Dwarf(t)

NATURE AND PROCEDURE OF THE EXPERIMENTS

Mendel had very clear concepts of what he was doing and what requirements were necessary if he had to arrive at the conclusions accurately and successfully. He realised the necessity of:

- Using pure breeding parent plants;
- Considering only one character at a time during the breeding experiments;
- Always keeping the generations separate;
- Using statistical and mathematical principles to compute the obtained results properly. Mendel conducted his experiments in three stages.

Stage-1 It involved selecting a pair of parents with contrasting characters and obtaining each parent plant in pure condition,i.e,. breeding true for the characters selected.

For example, Mendel ensured that the plant selected for round seeds produced only round seeds on self-fertilization and the plant selected for wrinkled seeds produced only wrinkled seeds. Such pure, true-breeding parents were obtained by Mendel through repeated self fertilizing, generation after generation.

Stage -2 It involved crossing of the selected parents. Of the pair, one plant was used as the male parent and the other as the female parent. Pollen from the male was dusted on the stigma of the female parent for cross-pollination. Mendel conducted reciprocal crosses also.

For example: in one cross, the round seeded variety was used as the male parent and the wrinkled seeded variety as the female parent, while in the reciprocal cross of the same parents, the wrinkled variety was used as the male parent and the round seeded variety as the female parent. This may be represented as follows:

	Male Parent	***Female Parent***
Cross I	Rounded seeded	Wrinkled seeded
Reciprocal Cross	Wrinkledd seeded	Rounded seeded

Such a cross between two parents representing contrasting forms of a single character is called monohybrid cross and the offspring is called a hybrid. The hybrid represents the first filial generation or F1 generation.

Stage - 3 In the third stage, Mendel allowed each F_1 hybrid to self-pollinate and produce the next, i.e., Second Filial generation or F2 generation.

Mendel conducted a similar type of hybridization experiment separately for each of the seven pairs. He meticulously maintained a complete record of the actual number of each type of offspring in every generation (i.e., data of qualitative as well as quantitative results).

Mendel observed that in each of these crosses, all the F_1 hybrids resembled only one parent, while the character of the other parent was not seen in the F_1 hybrid. For example, in a cross between round X wrinkled type, the F_1 were all round seeds only. The wrinkled character was not seen. The character which appears in the F_1 hybrid was termed dominant and the other as recessive by Mendel. Thus, in each of the seven pairs, one form is dominant and the other is recessive. The F_2 progeny showed presence of both parental forms which always appeared in the ratio of 3 dominants: 1 recessive. This 3:1 F_2 ratio was termed as the monohybrid ratio.

REASONS FOR MENDEL'S SUCCESS:

- Mendel concentrated on the results of one trait at a time.
- When the behaviour of one trait was established, only then he considered two characters together.
- He conducted a large number of crosses (2000-3000) to eliminate the chance factor and to obtain a valid and accurate explanation
- Most of all, he actually counted the number of offsprings of each category and maintained accurate records for each generation in each experiment.
- However, luck played a major role in his success (though Mendel did not know this fact) in the selection of pea plants as well as in the selection of those particular seven pairs of contrasting characters. Because, luckily for Mendel, in each pair, one form of the character is completely dominant over the other. Moreover, these seven characters are present on seven different chromosomes in the pea plant. However, Mendel was unaware of this fact.

- In nature, there are several instances of incomplete dominance also. In such cases, the hybrid shows haracter intermediate between the two parents. For example, in snapdragon, when a parent plant with red flowers is crossed with another having white flowers, the F_1 hybrids show pink flowers.

(*Parents*) Red Flower X White Flower

↓

(F_1 *Hybrid*) Pink Flowers

LAW OF DOMINANCE

On the basis of his experiments and observations, Mendel concluded that the characters from parents to offsprings are transmitted in the form of some kind of hereditary units or particles called factors (now called genes). These factors determine the characters in the individual. For each factor representing a character there are two alternate forms called alleles. Mendel represented the allelic factors by appropriate alphabets. In each of the seven pairs selected by him, the dominant allele was represented by the capital alphabet and its recessive allele by the same but smaller alphabet (e.g., Round seed = R, Wrinkled seed = r)

LAW OF DOMINANCE

In a hybrid union, the allele which expresses itself phenotypically is the dominant allele while the other allele which fails to express itself phenotypically is the recessive allele. The hybrid individual shows phenotypically only the dominant character.

The law of dominance is often described as Mendel's first law of inheritance. Dominance is seen in various characters in many plants and animals. But this is not of universal occurrence. There are many cases where dominance is incomplete or absent.

MONOHYBRID RATIO

The phenotypic ratio of different types of individuals

occurring in the F2 generation of the monohybrid cross is called the monohybrid ratio. In the Mendelian monohybrid experiments, this ratio was always 3:1(i.e., 75% is dominant and 25% is recessive). For example, for one of his monohybrid crosses, Mendel selected true breeding homozygous parents showing contrasting characters for the height of the plant. He performed the experiment in three stages as described. The result obtained is shown in below.

Table: Showing dominant and recessive characters in some plants and animals

	Examples	Dominant	Recessive	Appearance of F1 hybrids
PLANTS	Sorghum	(i) Pearly grain	Chalky grain	Pearly
		(ii) Awnless	Awned	Awnless
	Maize	Full Endosperm	Shrunken endosperm	Full
	Rice	Starchy endosperm	Glutinous endosperm	Starchy
ANIMALS	Rabbit	Black coat	white coat	Black
	Mice	Normal body size	dwarf	Normal
	Man	(i) Brown eyes	blue eyes	Brown
		(ii) Short stature	tall	short

Each sexually reproducing diploid organism possesses two 'factors' (genes) for each character; one factor is received (inherited) from male parent and the other factor is inherited from the female parent.

These two factors for a particular character are called alleles or allelomorphs. When an offspring receives identical alleles from both parents, it is called homozygous, pure or true breeding for the character. On the other hand, when the offspring receives dissimilar alleles from two parents, it is called heterozygous, impure or a hybrid for that character.

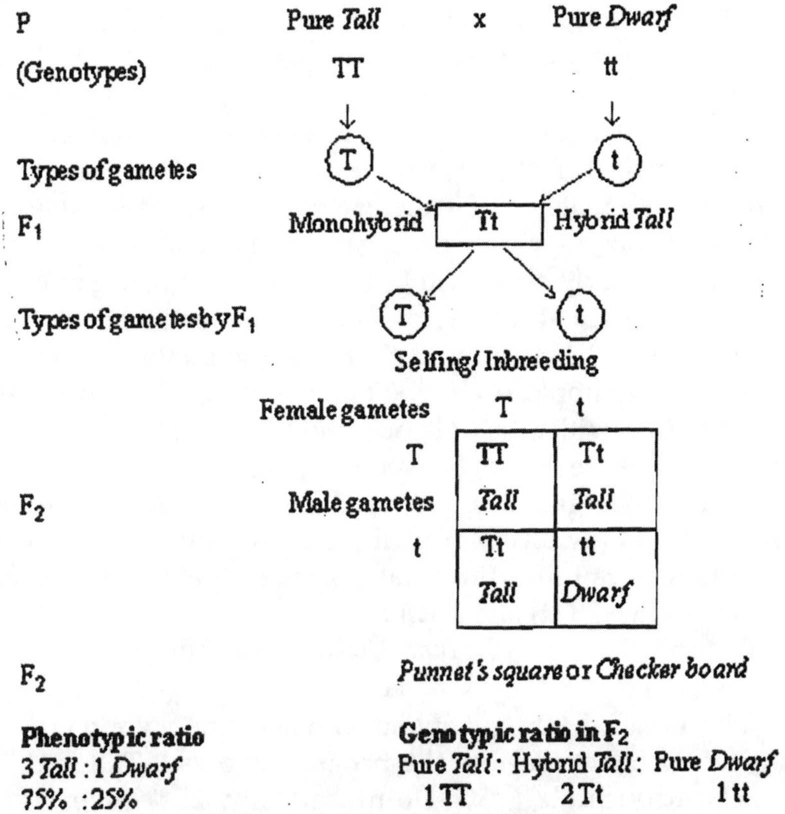

Fig. Tall x Dwarf Monohybrid cross showing the result obtained by Mendel.

The pure tall is crossed with the pure dwarf parent. When the diploid individual (having both the alleles/factors) produces gametes, each gamete receives only one of the two factors/alleles of a character. No gamete receives both the alleles of a character. Thus, pure tall parent produces only one type of gametes, i.e. all the gametes possess only (T) factor for tallness. Similarly, all gametes produced by pure dwarf are of one type only and possess (t) factor.

The fusion of (T) and (t) gametes (fertilization) results in the F1 offspring with (Tt) genotype. It is heterozygous or a hybrid. Its phenotype (external appearance) is tall because the factor for tallness (T) is dominant and expresses itself. The factor for dwarfness (t) is present in F1 hybrid but, being recessive, does not express itself (remains hidden).

Mendel allowed hybrids to self-fertilize or inbreed to raise F2 generation. The F1 hybrid has dissimilar alleles (Tt). Therefore, it will produce two types of gametes in equal number i.e. 50% gametes will have (T) factor and remaining 50% will have (t) factor. Since the pea flower is bisexual, it produces both male and female gametes. Thus, the F1 hybrid will produce two types of male gametes (T) and (t) in equal numbers. Similarly, there will be two types of female gametes (T) and (t) in equal numbers. During self fertilization, the fusion between these male and female gametes occurs at random. For example, each type of male gamete has an equal chance to fuse with either (T) or (t) female gametes and vice-versa. This chance fusion, between two types of male and two types of female gametes will produce a maximum of four combinations (genotypes) in the F2 progeny. These four combinations fall into three categories of the genotypes as follows: 1 (TT), 2 (Tt) and 1 (tt) i.e.

1 Pure tall:	2 Hybrid tall:	1 Pure dwarf
(TT)	2(Tt)	(tt)

This is called 1:2:1 genotypic ratio of a monohybrid cross. However, phenotypically, the progeny shows 3 Tall and 1 Dwarf individuals (75% Dominant and 25% recessive characters) or 3:1 ratio. This is called monohybrid ratio or phenotypic ratio of a monohybrid cross.

The result indicates that even though the recessive character was not seen in hybrid, it was present there and reappeared in pure form in 25% individuals of the progeny. This result also enabled Mendel to conclude that the two factors (alleles) come together in the hybrid but do not mix or fuse with each other. They simply remain together without diluting or contaminating each other. In other words, factors maintain their purity. As the gamete always receives only one factor (alleles) for a trait, it is always pure for the character. This is called purity of gametes.

LAW OF SEGREGATION

Mendel performed monohybrid crosses separately for each of the seven pairs of contrasting characters. In each such cross, he got similar results as described for the Tall x Dwarf cross. On the basis of the results obtained for the monohybrid crosses, Mendel formulated the law of segregation, also called Mendel's second law of heredity.

The law states that when a pair of alleles is brought together in hybrid union, the members of the allelic pair remain together without mixing, diluting or altering each other and separate or segregate from each other when the hybrid forms gametes.

Thus, according to this law, when the hybrid tall (Tt) in the above experiment forms gametes, the factors T and t shall separate (segregate) and enter different gametes. As a result, the hybrid shall form two types of gametes; those with (T) factor and an equal number with (t) factor. Since each gamete will be pure for tallness or for dwarfness, the law is also known as the law of the purity of gametes.

DIHIBRID RATIO

Mendel established the law of segregation through monohybrid crosses involving only one pair of alleles at a time. However, in an organism, so many characters are present together and each character is represented by a pair of alleles. So, Mendel wanted to know, "Whether one pair of alleles affects or influences the inheritance pattern of other pairs of alleles in the organism or, each pair is inherited independently as if in a monohybrid cross?" To find the answers to these questions, Mendel performed dihybrid crosses.

A cross between two pure, true breeding parents in which the inheritance pattern of two allelomorphic pairs is considered (studied) simultaneously is called a dihybrid cross. The phenotypic ratio obtained in the F2 generation of a dihybrid cross is called the dihybrid ratio.

A dihybrid is an individual which is double heterozygous (i.e. heterozygous for two pairs of alleles).

Mendel's Dihybrid Cross

Mendel considered two characters in the pea plants simultaneously, e.g. cotyledon colour (yellow / green) and seed shape (round / wrinkled). He selected one variety of pea which was pure (true breeding) for yellow round seeds and crossed it with another variety pure for green wrinkled seeds. All the F_1 of this cross were yellow round seeds (green and wrinkled characters did not appear in F_1 hybrids). Mendel anticipated this because, from the earlier monohybrid experiments he knew that yellow was dominant over green and round was dominant over wrinkled.

P Yellow Round X Green Wrinkled
↓
F Yellow Round

Similarly, a cross between yellow wrinkled and green round also produced only yellow round seeds in F_1

P Yellow Wrinkled X Green Round
↓
F Yellow Round

Moreover, the reciprocal crosses (interchanging male and female parents) also gave the same results. Further, when the F_1 dihybrids were self-pollinated or inbred, the F_2 generation was always the same, e.g.

P (yellow Round x Green Wrinkled) OR (Yellow Wrinkled x Green Round)
F → Yellow Round ←
↓
Yellow Round 9 | Yellow Wrinkled 3 | Green Round 3 | Green Wrinkled 1

The analysis of F_2 progeny showed four different kinds of phenotypes. These were:

- Yellow round
- Yellow Wrinkled
- Green round and
- Green Wrinkled in the ratio of 9:3:3:1 respectively.

It will be seen that out of these four types, two show the same combinations as the parents whereas the remaining two are new combinations (recombinants). The phenotypic ratio of 9:3:3:1 in the F_2 progeny of a dihybrid cross is called the dihybrid ratio. Same results were obtained by Mendel using other pairs of alleles in different combinations. On the basis of these experiments and their results, Mendel formulated the law of independent assortment of characters and explained it as follows.

LAW OF INDEPENDENT ASSORTMENT

"When a dihybrid (or a polyhybrid) forms gametes, (i) each gamete receives one allele from each allelic pair and (ii) the assortment of the alleles of different traits during the gamete formation is totally independent of their original combinations in the parents. In other words, each allele of any one pair is free to combine with any allele from each of the remaining pairs during the formation for the gamete. This is known as the Law of Independent Assortment of characters. It is also referred to as Mendel's third law of heredity.

Explanation of the Law of Independent Assortment

The principle of independent assortment was explained by Mendel with the help of a dihybrid cross involving characters of cotyledon colour (yellow / round) and seed shape (round / wrinkled). Mendel crossed a true breeding variety of pea having yellow cotyledons (YY) and round seeds (RR) with another true breeding variety having green cotyledons (YY) and wrinkled seeds (RR).

Thus, the yellow round parent has the genotype (YYRR) and the green wrinkled parent (YYRR). Since each parent is homozygous for both characters (colour and shape), each will produce only one type of gametes. The (YYRR) parent will produce all (YR) type gametes and the (YYRR) will produce all (YR) type gametes. All F_1 dihybrids resulting from the fusion of these gametes would be double heterozygous with (YYRR) genotype and appear yellow round. This indicated that in the dihybrid cross also in each pair, the alleles behaved exactly in the same way as in the monohybrid cross. Both the dominants (Y and R) expressed themselves in F_1 while both the recessive alleles (Y and R) remained hidden.

Types of Gametes Formed by F_1 dihybrid

During gamete formation by the F_1 dihybrid, the alleles in both pairs Y-y and R-r first segregate from each other (Law of segregation). Each pair segregates independently of the pair.

Then the alleles enter the gametes. A gamete can receive only one allele from each pair, i.e. Y or y and R or r. Similarly, a gamete that receives a factor (gene) for colour must also receive factor for shape (a factor for every character must be present in each gamete). Thus, a gamete that receives Y for colour may receive R or r for shape.

This would result in (YR) and (Yr) types of gametes. Similarly, a gamete that receives y for colour may receive R or r for shape. This would give (yR) and (yr) types of gametes. In other words, the F_1 dihybrid would produce four types of gametes (YR), (Yr), (yR) and (yr) in equal proportions. This is the principle of independent assortment of characters. There will be four types of male gametes and four types of female gametes formed by the F_1 dihybrid.

During self-fertilization or inbreeding of the F_1 dihybrids to produce an F2 generation, these male and female gametes can form maximum to dihybrid unions as shown in the Punnet's Checker-board. These can be grouped into four kinds on the basis of phenotypic appearance. i.e. yellow round, yellow wrinkled, green round and green wrinkled in the ratio of 9:3:3:1 respectively. This is called the Phenotypic dihybrid ratio.

The 16 squares of the checker board are serially numbered for convenience. The squares represent the 16 possible combinations of the gametes which might result. The genotypes and the phenotypes of the F_2 offspring are shown in the sixteen squares. A count of these squares shows the four kinds of phenotypes and their ratio in the F_2 generation.

Phenotypes	Square numbers		Total
Yellow Round	1,2,3,4,5,7,9,10,13	...	9
Yellow Wrinkled	6,8,14	...	3
Green Round	11,12,15	...	3
Green Wrinkled	16	...	1

In the actual dihybrid cross of the above type, Mendel obtained a total of 556 seeds in the F2 generation. The actual number of seeds of each type was as follows.

This gives approximately 9:3:3:1 ratio. Thus the actual

results agree completely with the theoretical expectations.

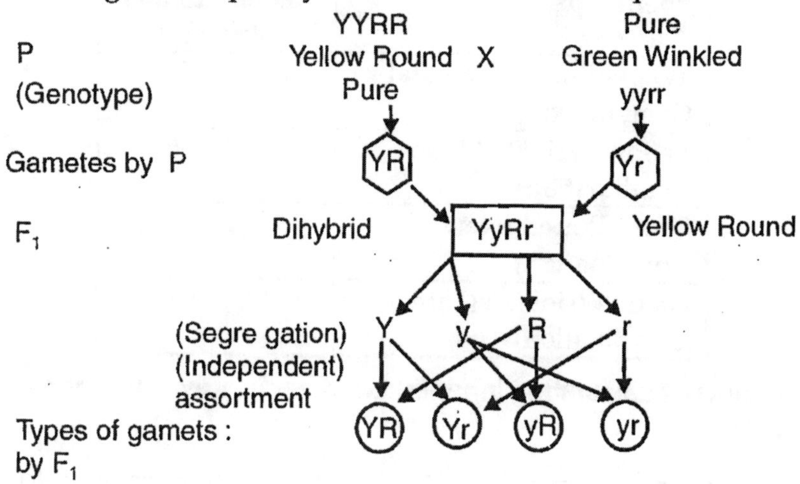

Selfing of F_1 dihybrids to produce F_2 generation F_1 Female Gamets

	YR	Yr	yR	yr
YR	1 YYRR Yellow Round	2 YYRr Yellow Round	3 YyRR Yellow Round	4 YyRr Yellow Round
Yr	5 YYRr Yellow Round	6 YYrr Yellow Wrinkled	7 YyRr Yellow Round	8 Yyrr Yellow Wrinkled
yR	9 YyRR Yellow Round	10 YyRr Yellow Round	11 yyRR Green Round	12 YyRr Green Round
yr	13 YyRr Yellow Round	14 Yyrr Yellow Wrinkled	15 YyRr Green Round	16 yyrr Green Wrinkled

Phenotype	Actual number obtained
Yellow Round (Parental Combination)	315/556
Yellow Wrinkled (New Combination)	108/556
Green Round (New Combination)	101/556
Green Wrinkled (Parental Combination)	32/556

Method of calculating inheritance of each individual character (gene)

Multiplication rule states that the probability of two or more independent events occurring together is the product of their individual probabilities. In the dihybrid cross, the gene for shape (round/wrinkled) and the gene for colour (yellow/green) are present together but are sorted independently (law of independent assortment).

Calculation of F_2 Phenotypes in a Dihybrid Cross

The multiplication rule applies to the probabilities of the phenotypes. For each gene, the monohybrid phenotypic ratio is 3:1 or ¾: ¼ Thus, for the colour gene the monohybrid phenotypic the ratio is 3 yellow: 1 green or ¾ yellow : ¼ green. And, for the shape gene, it is 3 Round:1 wrinkled or ¾ Round: ¼ wrinkled.

$$(3/4 : 1/4) \times (3/4 : 1/4) = \frac{9}{16} : \frac{3}{16} : \frac{3}{16} : \frac{1}{16}$$

Therefore, in the dihybrid cross, the total probability of the F_1 phenotypes will be the product of the two monohybrid ratios.

This is calculated as follows:

Gene for Shape	Gene for colour	Probabilities of Phenotypes

TEST CROSS OR BACK CROSS

Test cross is a simple method devised by Mendel to verify

the genotype of the F_1 hybrid. When the F_1 hybrid is crossed with the homozygous recessive parent, it is called a test cross. Since, the F_1 is crossed back with one of the parents, it is also called a back cross.

Test cross is also used for checking the correctness of Mendel's law of segregation (using a monohybrid test cross) and the law of independent assortment of characters (using a dihybrid test cross). For example,

$$\frac{3}{4}\text{ round} \times \begin{cases} \frac{3}{4}\text{ yellow} = \frac{9}{16}\text{ round yellow} \\ \frac{1}{4}\text{ green} = \frac{3}{16}\text{ rounc green} \end{cases}$$

$$\frac{1}{4}\text{ wrinkled} \times \begin{cases} \frac{3}{4}\text{ yellow} = \frac{3}{16}\text{ wrinkled yellow} \\ \frac{1}{4}\text{ green} = \frac{1}{16}\text{ wrinkledgreen} \end{cases}$$

Monohybrid Test Cross

In a monohybrid cross of Tall x Dwarf, the F_1 are all tall. Let us see what happens when this F_1tall is test crossed with the homozygous recessive parent i.e. Dwarf with (tt) genotype. We aim to check and verify two things (i) determine the genotype of F_1 Tall and (ii) check the correctness of the law of segregation.

Thus, F_1 tall is back crossed (test crossed) with the homozygous recessive dwarf parent and the progeny of the cross examined. We know that the recessive dwarf with (tt) genotype will produce only one type of gametes (i.e., all with (t) only). However, as regards the F_1 tall, there can be two possibilities:

- If the F_1 tall is homozygous with (TT) genotype, it shall produce only one type of gametes (i.e. all with (T) only). As a result, the progeny of the cross should be all tall. (T) x (t) = (Tt) Tall.
- If, on the other hand, the F_1 tall is heterozygous with (Tt) genotype and, if Mendel's law of segregation of characters is correct, then the F_1 should produce two types of gametes, (T) and (t) in equal proportion. The

recessive dwarf parent produces only (T) type of gametes. The resulting test cross progeny should be 50% Tall and 50% Dwarf or in 1:1 ratio as follows:

Test cross	Heterozygous F_1 *Tall*	X	Homozygous *Dwarf* parent
Genotypes	(Tt)		(tt)
Types of gametes	(T) (t)		(t)

Test cross progeny		(T)	(t)	
	(t)	Tt *Tall*	tt *Dwarf*	1 *Tall* : 1 *Dwarf* 50% 50%

The results obtained in the actual test cross experiments completely agree with the theoretical expectations. Thus it is proved that F_1 Tall is a heterozygous dominant (monohybrid) with (Tt) genotype and that the alleles segregate during gamete formation.

Dihybrid test Cross

In the dihybrid cross of yellow round X green wrinkled the F_1, dihybrid is a double heterozygous yellow round with (YyRr) genotype. It is testcrossed with the double homozygous recessive parent,(i.e. green round with (yyrr) genotype). We aim to test the correctness of Mendel's law of independent assortment of characters. If the principle of independent assortment is correct, then as Mendel explained, the (YyRr) dihybrid should produce four types of gametes: YR, Yr, yR and yr in equal proportions. The recessive parent (yyrr) shall produce only one type of gametes i.e. yr type. It is expected that the maximum possible chance combinations between these gametes should produce four kinds of phenotypes in the ratio of 1:1:1:1 as follows.

The actual numbers of these four types of offspring obtained in one of Mendel's test crosses were: Yellow round = 55, Yellow Wrinkled = 49, Green Round = 51 and Green Wrinkled = 52. This gives an approximate 1:1:1:1 ratio as per the theoretical expectations and thus, confirms the law of independent assortment of characters.

	[Double heterozygous F_1 Dihybrid]	[Double heterozygous recessive parent]
Test cross	Yellow Round	Green Wrinkled
Genotypes	YyRr	yyrr
Types of Gametes	YR Yr yR yr	yr

	YR	Yr	yR	yr
Test cross yr Progeny	YyRr yellow Round	Yyrr Yellow Wrinkled	yyRr Green Round	yyrr Green Wrinkled
	1	1	1	1
	25%	: 25%	: 25%	: 25%

Significance of Back Cross in Plant Breeding

- It is a rapid method of improving a variety of crop.
- It is an easier and quicker method of obtaining a desirable trait in a pure homozygous condition (true breeding trait).
- Back cross is also use frequently in hybridization experiments because of its simple ratio. Moreover, this needs the analysis of fewer progeny.

THE CONCEPT OF "FACTOR"

The period in which Mendel conducted his experiments and drew his conclusions about the principles of heredity and variations, the world was totally ignorant about 'genes' (as the units of heredity) and the 'chromosomes' (as the carriers of hereditary units). Mendel believed that some definite hereditary units or particles represent the characters. He called these hereditary units 'factors'.

Thus, tallness is due to a factor for tallness and dwarfness is due to the factor for dwarfness. Since tallness or dwarfness are two expressions or forms of the same character,(i.e. height), it is also understood that their factors also are two forms of the same gene. Such two factors representing different forms of the same character (or its gene) are described as alleles.

We now know these 'factors' as genes. The term 'gene'

was first coined by Johansen in 1909. Chromosomes were discovered by Strasburger in 1875. Now it is established that genes are definite units of the DNA molecule of each chromosome. Genes are present in a linear fashion on the chromosomes. Each chromosome carries several genes 'linked' together in 'linkage groups.' Hence, genes are not inherited individually as was thought by Mendel but are passed on to the next generation as 'linkage groups' (i.e., as whole chromosomes).

This fact alters Mendel's principle of independent assortment. However, the principle of segregation remains unchanged even today. The studies on genetics have shown that genes influence certain biochemical reactions which in turn are responsible for the production of characters in organisms. However, a single character is not (always) produced by a single gene as believed by Mendel. On the contrary, several genes influence the same basic biochemical reaction and are responsible for the expression of a single character.

Chapter 5

Hormonal Coordination

All animals contain chemical components produced by exocrine or endocrine glands. Exocrine glands release their chemicals into ducts. The endocrine glands release their secretions, the hormones, directly into the blood-stream. These hormones are then carried to other parts of the body (target organs), where in extremely minute quantities, they elicit cellular responses. E.H.Starling suggested the name 'hormone' (Greek meaning "to stimulate" or "to excite") for these chemical messengers or transmitters.

Hormones are specifically-acting organic compounds with varying chemical compositions, usually steroids, proteins, peptides or amino acids. Based on their general roles, hormones are metabolic (stimulate or retard metabolic activities), trophic (regulate rate and secretion of other endocrine glands) and morphogenetic (affect rate and development of various parts). Hormones are an additional means of coordination and communication. Together with the nervous system, the endocrine system forms a combined neuro-endocrine system. The hormone pathway is by the blood stream, while nervous pathway is by the neuron-reflex arc.

PITUITARY GLAND

Position

The pituitary gland (hypophysis) is about the size of a large pea, located below the brain, behind the nasal cavity on the floor of the cranium. It is attached by a stalk to the hypothalamus portion of the brain on the undersurface of the cerebrum. It lies in the sphenoid bone of the skull and is closely invested by a connective tissue capsule.

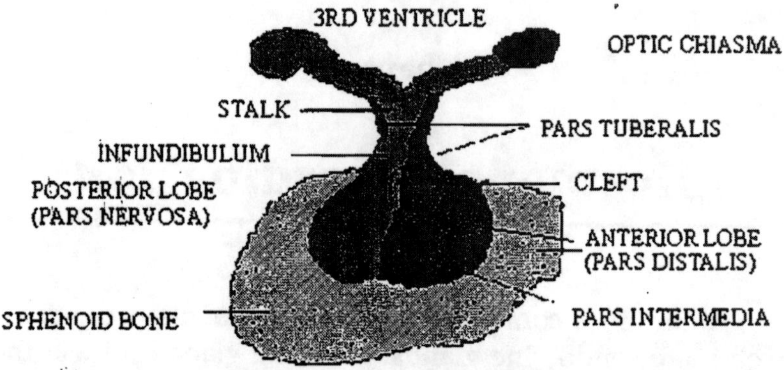

Fig. Position of Pituitary gland

The pituitary has been called the "master endocrine gland" because of its control over several other endocrine glands. However, the body's real master gland is the hypothalamus of the brain which controls the secretions of the anterior pituitary (by releasing and inhibiting hormones). Also, the pituitary itself is in turn controlled by feedback from other glands.

Structure

The sagittal section shows that the pituitary consists of two main lobes, the nterior lobe (pars distalis) and the posterior lobe (pars nervosa). The small intermediate lobe (pars intermediate) is marked by a cleft (in children) or cystic spaces (in adults). The part of the anterior lobe which extends upwards as part of pituitary stalk is called the pars tuberalis. The pars anterior, pars tuberalis and pars intermedia form the adenohypophysis while the pars nervosa and the pituitary stalk (infundibulum) form the neurohypophysis.

The anterior lobe produces six different hormones. Of these, two work directly on body tissue (somatotrophic) and the other four control the action of other endocrine organs. The middle lobe (pars intermedia) secretes only one hormone, intermedin, which controls skin colours in lower vertebrates, but is vestigial in mammals, and has no function in humans. The posterior lobe serves as a storage and release point for two hormones secreted by the hypothalamus (neurosecretory cells) and carried to the lobe through a connecting duct (i.e., the neurohypophysis).

Histology

Histological structure of the pituitary shows different types of cells in three different parts of the pituitary. The cells are classified as chromophobes or chromphils by the staining reaction of their cytoplasmic granules. In the anterior lobe, chromophobe cells are smaller, nuclei clumped together and are clustered near the centre. Chromophobe cells stain poorly. The chromophil cells stain red or purple blue, contain secretory granules and are distributed at the periphery.

The nuclei differ in size and in staining intensity. They are of two types, oxyphilic cells and basophilic cells. Blood vessels and lymph spaces are also seen. The intermediate lobe is poorly developed in man and contains smaller cells than the anterior lobe. The cells in the posterior lobe show neurosecretory granules, prominent nuclei and polygonal supporting neuroglial cells. This is because the posterior lobe stores the hormones secreted by the neurosecretory cells of the hypothalamus.

Pituitary Hormones

Anterior lobe.

The anterior lobe of the pituitary secretes six hormones.

- Thyroid stimulating hormone (TSH), or thyrotropin, stimulates the secretion of thyroxin from the thyroid glands.
- Somototropic or growth hormone controls growth of tissue, bone, muscles and internal organs, and also influences metabolic processes. Hypersecretion in early life gives rise to a condition called gigantism, and after the growth period in adult life it results in acromegaly. In acromegaly, excess secretion appears after long bones have ossified and the growth continues only in certain bones, resulting in

lengthening of forelimbs, hindlimbs, and enlarged jaw and face. Hypofunction during growth years results in pituitary infantilism, a kind of dwarfism, in which an adult is not more than three or four feet tall, and is usually sexually immature.

- Follicle stimulating hormone (FSH) stimulates the ovaries and testes. In females, it controls the development of grafian follicles (ova). In males, FSH influences spermatogenesis, i.e. production of sperm by the seminiferous tubules of the testes.
- Adrenocorticotropic hormone (ACTH) stimulates the adrenal cortex and regulates the secretion from the adrenal glands.
- Luteinizing hormone (LH) along with FSH, is responsible for the growth of ovarian follicles, release of mature ova, formation of the corpus luteum, production of progesterone and regulation of menstrual cycle in females. In males, LH stimulates the testes to secrete testosterone or ICSH- interestitial cell stimulating hormone.
- Prolactin initiates the secretion of milk by the mammary glands after the birth of a baby. In the absence of prolactin, milk secretion stops.

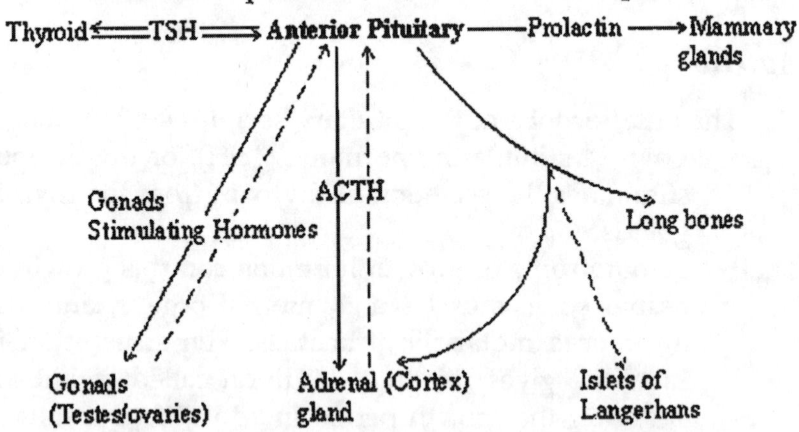

Fig. Interaction Among Pituitary and other Endocrine Glands
(– – – –Responsible for feed back or inhibition)
(________Pituitary Hormones that stimulate other glands)

Intermediate Lobe

Intermediate lobe secretes only one hormone, Melanophore Stimulating Hormone (MSH), or intermedin, which controls skin pigmentation of many vertebrates. It makes the colour of the skin dark. However, it has no function in humans.

Posterior Lobe

- Oxytocin causes contraction of the uterus in females during childbirth and regulates lactation (release of milk).
- Antidiuretic hormone (ADH) or Vasopressin stimulates the kidneys to reabsorb more water, preventing excessive water loss by urination. Chronic deficiency of ADH results in diabetes insipidus (polyuria), in which the patient feels extremely thirsty and passes a large amount of dilute urine.

THYROID GLAND

Position and Structure

All vertebrates possess a pair of thyroid glands, located in the neck just below the larynx or Adams apple. In humans, the thyroid consists of two lobes (H-shaped) that lie on either side of the trachea. The two lobes are connected by a narrow isthmus that passes in front of the trachea. The gland varies in size with the difference in sexual development, diet and age.

Histology

The thyroid gland consists of a large number of round or oval follicles surrounded by connective tissue with a large number of blood vessels. Each follicle is lined by one-cell thick cuboidal epithelial cells. The cavities of the follicles are filled with viscous protein material called colloid. The thyroid hormone, thyroxine contains iodine atoms. It is not stored in the cells of the thyroid, but in the colloid filling the follicles. The role of thyroid cells is to strain iodine out of the blood. This is then incorporated into the protein thyroglobulin, which is then hydrolyzed into the active hormone thyroxine.

Thyroid Hormones

Thyroid glands produce a hormone called thyroxine. The secretion of thyroxin is controlled by TSH produced by the pituitary. The hyposecretion of thyroxine in a child results in cretinism. In this, retarded growth and development (dwarfism), a protruding abdomen, mental retardation, puffy skin and low metabolic rate occurs.

In adults hypofunction results in myxedema; the symptoms are a low metabolic rate, reduction in mental and physical activity, increase in weight, puffy skin, a decrease in heart beat and body temperature, and loss of hair. When hypofunction results due to lack of iodine in diet, enlargement of the thyroid results in a condition called simple goiter. In an effort to capture more iodine from the blood, the gland increases follicles and grows disproportionately. Hypofunction is treated with a supply of iodized salt, sea food, and by surgery to remove excess of thyroid tissue.

In amphibians, thyroxine plays a critical role during metamorphosis. If thyroid function of an embryo or young tadpole is inhibited (for example by excising the thyroid gland), then the animal remains a tadpole permanently. Conversely, if a young tadpole is given an excess of thyroxin, the larva metamorphoses prematurely into a tiny froglet. Hyperfunction results in an increased metabolic rate up to (40%), profuse sweating, increased food intake but loss of weight, high blood pressure, nervous tension and muscular weakness.

Some patients with hyperthyroidism have protruding eyeballs, a condition called exophthalmos. The swelling of thyroid due to hypersecretion produces exophthalmic goiter. Another hormone produced by the thyroid gland is calcitonin. It maintains the blood calcium level. Calcitonin is secreted by the thyroid when there is a high level of calcium in the blood. The excess Calcium is then decreased and deposited in bones.

NERVOUS CO-ORDINATION

The behaviour of an animal in its environment to maintain itself depends on the coordination of its organs systems. Without coordination of various organ systems, various physiological processes would work in a haphazard way, without linking together activities. The linking together in time and space of various activities of an animal is called coordination. Coordination is brought about by the nervous system and sense organs, and by means of chemical substances (hormones) secreted by the endocrine glands.

During the course of evolution, nervous systems have become more complex. This is largely because animals have become larger and more mobile, requiring more neurons than a sedentary animal. The most fundamental function of a nervous system is:

- To receive a stimulus
- Transmission of a stimulus to a central "brain",
- Interpretation and analysis of the stimulus
- Proper response by an effector.

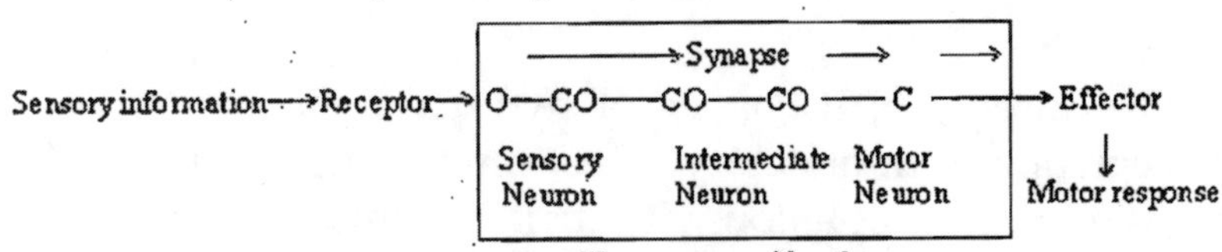

Nervous co-ordination

With the evolution of a complex nervous system and bilateral symmetry, cephalization (formation of a head) has taken place, and the ganglia in the head became large enough to be called a brain, which is the main nervous control centre of the body.

The Human Nervous System: The human nervous system is divided into two interrelated parts:

- The central nervous system (brain and spinal cord).
- The peripheral nervous system (nerves and ganglia).

Divisions of the human nervous system

Divisions of the human nervous system

Central Nervous System	Peripheral Nervous System		
BRAIN	Cranial Nerves (12 pains)	Somatic Nervous System (Vohuntary)	
SPINAL CORD	Spinal Nerves (31 Pairs)	Autonomic Nervous System (Invohuntary)	Sympathetic Nervous System Parasym pathetic Nervous System

CENTRAL NERVOUS SYSTEM

BRAIN

The human brain weighs about 1.5kg (average 1350 gms or 3 pounds) and is enclosed in the cranial (skull) cavity. It is covered by 3 membranes, called meninges. The outer membrane, called dura mater ("tough mother") is a tough, protective covering formed of white fibrous tissue the middle delicate membrane is called arachnoid (web like), and the inner most transparent membrane is called pia mater ("soft mother"), containing blood vessels (nutritive in function). Between the bony and membranous coverings, there is cerebrospinal fluid, which acts as a protective internal cushion. The bony skull, meninges and the cerebrospinal fluid protect the brain from external injury and shocks.

Dorsal view

The Forebrain: It consist of two divisions, the anterior telencephalon with olfactory lobes, cerebral hemispheres and the posterior diencephalon with thalamus and hypothalamus. The cerebrum is the largest area of the brain and is divided by a longitudinal fissure into two cerebral hemispheres. Each hemisphere consists of a frontal lobe, parital lobe, occipital lobe and temporal lobe. It shows prominent fissures (deeper furrows) or sulci (shallow grooves) and gyri or convolutions (thin ridges between sulci).

Structurally, the hemispheres are made up of 3 portions:

- Cerebral cortex or grey matter,
- Cerebral tracts are bundles of axons located inside the brain and extend in three directions forming projection fibers, association fibers, and transverse fibers forming the corpus callosum,
- Basal nuclei or ganglia located deep within the cerebral hemispheres, connected to each other and with the thalamus and hypothlamus.

CEREBELLUM

The thalamus is a large, oval part located above the midbrain. All sensory impulses (except for olfactory sensations) entering the cerebrum must pass through the thalamus. The hypothalamus forms the floor and lower part of the brain formed by the thickening of the diencephalon. Part of the hypothalamus also extends downwards and contributes to the pituitary body. Here, the nervous and hormonal systems interact. It is concerned with different bodily activities, homeostasis, and is the centre for the parasympathetic nervous system. Internally, the downgrowth from the diencephalon forms the pituitary body, while the thin and vascular roof (anterior choroid plexus) secretes cerebrospinal fluid into the third ventricle.

The midbrain is a short, constricted mass mostly of white matter surrounding a central cavity. Ventrally, it connects the pons and cerebellum with the cerebral hemispheres by tracts of nerve fibers. In mammals, the analysis of vision has moved out of the midbrain and has become a function of the forebrain. The dorsal part of the midbrain has four small rounded projections which contain visual reflex (superior colliculi) and auditory reflex centres (inferior colliculi).

The hindbrain or rhombencephalon consists of two subdivisions, anterior metencephalon and a posterior myelencephalon. The hindbrain has 3 main parts:

- Cerebellum
- Medulla
- Pons.

The cerebellum occupies the lower posterior part of the skull cavity. It is the second largest part of the brain and consists of the outer layer of grey matter (cortex) and an inner core of white matter (medullary body). Its surface is also grooved, but the convolutions are less prominent than cerebral hemispheres. The cerebellum has two large lateral masses, the cerebellar hemispheres, and a central portion, the vermis. The internal white matter has short and long nerve tracts, which connect various portions of the cerebellum with each other, and with other parts of the brain and spinal cord.

The medulla (medulla oblongata), also called the "brainstem" is the anterior enlargement of the spinal cord, and consists of ascending and descending tracts of white matter. The vital nuclei in the medulla control vital reflex centres such an heart action, respiration, diameter of blood vessels, etc. The nonvital nuclei contain reflex centres for activities such as vomiting, coughing, swallowing, sneezing, etc.

The pons or pons varioli lies just above the medulla and is composed of white matter. It contains transverse tracts that conduct impulses between the cerebellum, medulla and cerebrum. In pons, the nerve tracts between brain and the spinal cord cross from the left side to the right side. Because of this crossover, the left side of the brain controls activities of the right side of the body and vice versa.

FUNCTIONS OF DIFFERENT PARTS OF THE BRAIN

- The cerebral hemispheres (cerebrum) perform (a) all mental functions, reason, will, memory and intelligence, and (b) many essential motor sensory and visceral activities. Certain functional areas of cerebral cortex like sensory, motor and association areas are mapped out after experimentation; these control motor, sensory and other activities. The portion of cerebrum that governs muscular movements is known as the motor areas; those involved in analysis of sensations are called sensory areas and those concerned with the higher faculties such as reasoning, will, understanding, memory, personality, ethics, etc., are called association areas.
- Basal ganglia or nuclei play a role in producing movement.
- The thalamus is associated with pain, pleasure and emotions. It functions as an arousal or alert mechanism of the body. It also has a role in producing complex reflex movements.
- The hypothalamus is the control centre for homeostasis, partial control of sleep and wakefulness, regulation of body temperature and food intake, and controls of metabolism and water balance through synthesis of ADH secreted by pituitary, regulation of autonomic activities, and control of various reproductive functions.

- The pituitary body is the master endocrine gland.
- The cerebellum coordinates the voluntary skeletal muscle activities and several other brain centres. It controls skeletal muscles related to maintenance of equilibrium, smooth, timed, precise and steady body movements.
- The medulla is the automatic control centre for the heart beat, breathing, swallowing, sneezing, etc. Sleep and loss of consciousness and cerebrum activities are controlled by tissues within the medulla.
- The pons serves as a reflex centre, regulates respiration, and serves as a conduction pathway between the spinal cord and the brain.

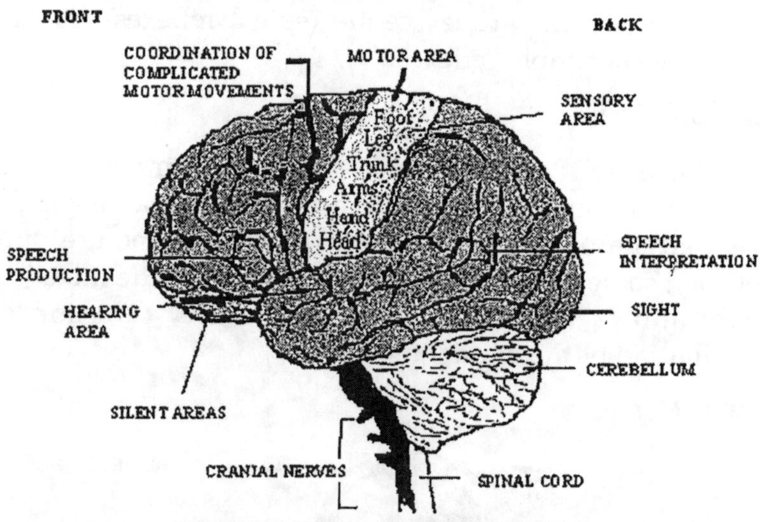

Fig. Functional areas of the Brain

THE SPINAL CORD

The spinal cord extends from the medulla or brain stem down to the second lumbar vertebra. It is surrounded and protected by the meninges (same as covering the brain) and the vertebrae of vertebral column. It is a hollow, oval shaped cylinder, tapering slightly as it descends. A transverse section of the spinal cord has an H-shaped region of grey matter. The grey matter contains nerve cells and non-myelinated fibers.

A small central canal containing cerebrospinal fluid runs

through the centre of the chord. Outside the grey matter is white matter comprised of numerous medullated nerve fibers running up (ascending) and down (descending) the cord or passing out to the spinal nerves. These fibers usually cross over from one part of the spinal cord to the other along their pathways to and from the brain. Thus, the right side of brain receives impulses that originate on the left side of the body and vice versa.

Functions

- The spinal chord serves as a pathway for the condition of impulses between the receptors and effector organs and the brain.
- It acts as a reflex centre (spinal reflexes) for many local simple reflexes.

Cranial Nerves

There are 12 pairs of cranial nerves in mammals (10 pairs in fishes and amphibians) and most of them lead away from the medulla oblongata. Of these 12 pairs, some are purely sensory, some are purely motor and the others are mixed. The table shows the name, number, nature, function and distribution of the cranial nerves.

Spinal Nerves

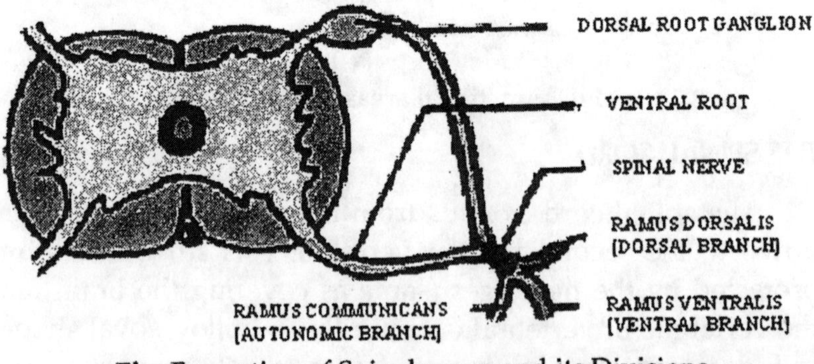

Fig. Formation of Spinal nerve and its Divisions

There are 8 cervical, 12 thoracic, 5 lumbar, 5 sacral and 1 coccygeal pairs of spinal nerves (31 pairs in all), which leave the spinal cord between adjacent vertebrae. Each spinal nerve is formed by the joining of two roots, a dorsal root (somatic and visceral sensory) and a ventral root (somatic and visceral motor), to form a mixed nerve.

CRANIAL NERVES OF MAN

No.	Name	Nature	Function	Distribution
1.	Olfactory	Sensory	Smell	From nose
2.	Optic	Sensory	Vision	From eye
3.	Oculomotor	Motor	Eye movement	To muscles of eye ball
4.	Trochlear	Motor	Eye movement	To muscles of eye ball
5.	Trigeminal	Mixed	Sensitive and jaw movement	From and to face, teeth, lips, tongue,jaws
6.	Abducens	Motor	Eye movement	To muscles of eye ball
7.	Facial	Mixed	Taste sensation, jaw movement	From taste buds, to sali vary glands and face
8.	Auditory	Sensory	Hearing and balance	From ear
9.	Glossophar yngeal	Mixed	Muscle movement and sensations	From and to pharynx, from taste buds Salivary glands
10.	Vagus	Mixed	Sensory to chest and abodomen	From and to visceral organs
11.	Spinal	Motor	Movement of shoulder muscles	To shoulder muscles
12.	Hypoglossa l	Motor	Movement of tongue	To tongue

The nerve soon splits into three branches, each containing

some sensory and some motor fibers. Ramus dorsalis serves the skin and muscles of the back, another innervates skin and muscles of the front of the body, and the third serves internal organs (autonomic branch). The cervical spinal nerves innervate diaphragm, skin and muscles of the neck and arms, the thoracic to skin, muscles of thorax, the lumbar nerves innerrate to skin and muscles of thigh, hip and lower leg, while the sacral and coccygeal nerves innervate skin and muscles of leg, foot, genitalia and abdomen.

After many spinal nerves emerge from the cord, they branch extensively to form networks of nerves called a plexus. Four of such plexuses are formed (i.e. cervical plexus in the neck region, brachial plexus in thorax, lumbar plexus in the abdomen and sacral plexus in the lumbo-sacral region).

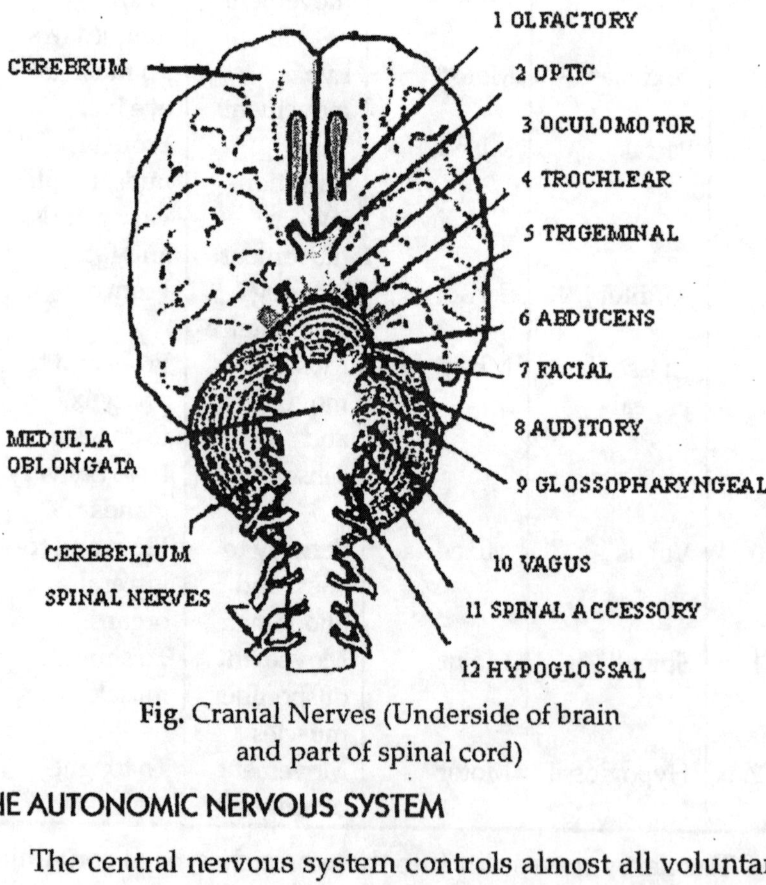

Fig. Cranial Nerves (Underside of brain and part of spinal cord)

THE AUTONOMIC NERVOUS SYSTEM

The central nervous system controls almost all voluntary

activities of the animal. The autonomic nervous system controls automatic activities which are free of will (involuntary such as heart beat, breathing, peristalsis etc.) Thus, this system deals with the internal environment of the body (homeostasis). The autonomic nervous system is connected with the central nervous system and is divided into two parts:

	Sympathetic Nervous System		**Parasympathetic Nervous System**
1	Contains nerve fibers on which there is no voluntary control. (Involuntary nerve fibers.)	1	--------- same ---------
2	Nerve fibers emerge from the brain or spinal cord.	2	--------- same ---------
3	The final nerve fibers supplying organ do not have thick myelin sheath, thus sometimes called non-myelinated fiber.	3	--------- same ---------
4	The ganglia lie close to the vertebrae/spinal cord.	4	The ganglia are embedded in the wall of effector itself.
5	Pre-ganglionic fibers are short.	5	Pre-ganglionic fibers are long.
6	Post-ganglionic fibers are long.	6	Post-ganglionic fibers are short.
7	Sympathetic fibers are generally activatory. (i.e. increase rate of heart beat.)	7	Parasympathetic fibers are generally inhibitory (i.e. decreases rate of heart beat).
8	Functions antagonistic to para-sympathetic fibers.	8	Functions antagonistic to sympathetic fibers..
9	Terminals of post-ganglionic fibers secrete adrenaline or epinephrin (adrenergic fibers).	9	Terminals of post ganglionic fibers release acteylcholine (cholinergic fibers.)

- Sympathetic nervous system
- Parasympathetic nervous system.

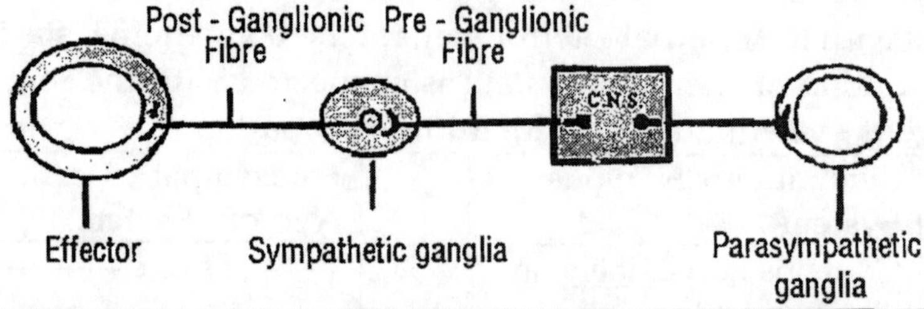

S.No.	Type	Function	Location
1	Mechanoreceptors	Sensitive to tactile stimuli : vibrations or pressure change and sound	Skin and ears
2	Thermoceptors	Sensitive to thermal or pressure changes and sound.	Skin
3	Chemoreceptors	Sensitive to concentrations of chemicals intensity and wave-length of light.	Tongue, olfactory chambers
4	Photoreceptors	Sensitive to change in in the internal and external environment	Eyes

RECEPTORS AND EFFECTORS

Sense organs make an animal aware of conditions or events in its own body and in the world around it. Information collected by sense organs is passed to the nervous system, which determines and initiates an appropriate response. The stimuli or the sensations are received by the nerves which are associated with special sensory cells called receptors. Impulses from the receptors reach the brain (Central Nervous System), where they are interpreted or analysed, following which a proper command is given to the effector organs. The effector organs are usually muscles or glands which contract or secrete chemicals producing suitable responses.

In higher animals, different types of cells (receptor cells) receive a variety of stimuli. These receptor cells are grouped together to form a sensitive structure known as sense organs. Generally, five senses are known in higher animals. These are, sense of touch, smell, taste, hearing and sight. A particular sense receptor or a sense organ is specialized and responds to only one kind of stimulus.

REFLEX ACTION - MECHANISM OF NERVOUS ACTION

The term reflex action refers to a rapid, automatic unlearned response to a stimulus. These actions are involuntary as opposed to voluntary actions, which are governed by willful or conscious control of the animal.

A familiar example of the reflex action in man is the "knee jerk" reflex or the "hand withdrawal" after pin prick. It is common understanding that if suddenly one is pricked by a sharp object such as a pin one immediately retracts from the object. In such actions, the impulses are carried by the sensory fibers from the receptor organ to the central nervous system and back to the effector organ. The path or the route through which the impulses travel from the receptor organ (via the Central Nervous System) is called a reflex arc. If the controlling centre of the reflex arc is located in the brain it is called cerebral reflex. If it is located in the spinal cord, it is called spinal reflex.

The reflex arcs are of two types:

- Somatic reflex arc involving effector organs located in (soma) body structures, e.g. skeletal muscles.
- Visceral reflex arc involves effectors located in the visceral organs; e.g. glands or smooth muscles.

Components of the Reflex Arcs

A simple reflex arc consists of only one sensory neuron and one motor neuron (monosynaptic reflex). But in a majority of reflexes, a sensory neuron influences several motor neurons through intermediate (association) neurons of the spinal cord. This forms a complex reflex arc, or polysynaptic reflexes. Some common examples of reflex actions are knee jerk reflex, contraction of iris diaphragm (i.e. narrowing of pupil in response of intense light) blinking of eye lids when aimed at, withdrawal of hand on pin prick, sneezing, and secretion of saliva at the sight or smell of food.

Somatic reflex arc	Visceral reflex arc
1. Receptor organ (e.g.) skin	1. Receptor organ (e.g. Epithelial lining of the gut)
2. Somatic sensory or afferent neuron (unipolar)	2. Visceral sensory or afferent neuron (unipolar)
3. Intermediate neuron (Multipolar)	3. Visceral motor of efferent (Pre-ganglionic)
4. Somatic motor or efferent neuron (Multipolar)	4. Visceral motor or efferent neuron (Post-ganglionic)
5. Effector organ (e.g. Skeletal muscles)	5. Effector organ (e.g. gland or smooth muscles)

In all such actions, the impulses are carried to and from the spinal cord In other words a reflex action is an involuntary action and there is no choice as to whether the action should take place or not. Many of the reflexes are innate, and are called unconditional or inherited reflexes. These reflexes are few in number and do not depend on a prior experience. Previous experiences establish conditioned or acquired reflexes.

They are much more in number and develop due to environmental changes. The conditioned reflex is one way in which animal and humans learn. In all reflex actions, the spinal cord forms the main path for the impulses, from and to the brain. Reflex actions have survival value.

LOCOMOTION

Movement is one of the most functional characteristics of animals. In most animals, movement or locomotion is brought about by special fibrils or muscular tissues. The basic components consist of the external body covering the supporting framework (skeleton) and the skeletal muscles, and provide the means for locomotion. These components also protect and support internal structures and provide body shape and form. Locomotion may occur in the form of walking, running, flying etc. This helps the animal to capture food, escape from predators, find its mate for reproduction, and many such activities which help in its survival.

The process of locomotion takes place on certain physical principles such as density, forces (thrust or pull), machines (levers in animal bodies), friction, buoyancy (upthrust in air or water) and stability which is important in a body at rest or in motion. In most terrestrial mammals, the hindlimbs provide the pushing force or thrust which moves the body forward by foot pressure against the ground. The foot remains in contact with the ground by friction. The hindlimb produces a power stroke through straightening of the limb by contraction of extensor muscles, and a recovery stroke by bending the limb with flexor muscles.

JOINTS - ARTICULATIONS BETWEEN BONES

A joint is that part of the endoskeleton where two or more bones meet. Joints between bones are classified into three main types on the basis of both structure and function, though these are, of course, related.

FIBROUS OR IMMOVABLE JOINTS (SYNARTHROSIS)

This includes all immovable joints in which the bones are united by a thin or dense layer of fibrous tissue. The cracks between the bones are called sutures. These joints are places of growth and remain open till the growth period is over, after which they tend to ossify. (Most sutures in marsupials remain open for life).

- Butt joint. Sutures are straight or have squared off edges. Butt joints can withstand compression, but not shear or bending. (e.g., two nasal bones).
- Lap joint. When the squared edges of bones overlap, it forms the lap joint. Lap joints are stronger, and unusual in animals.

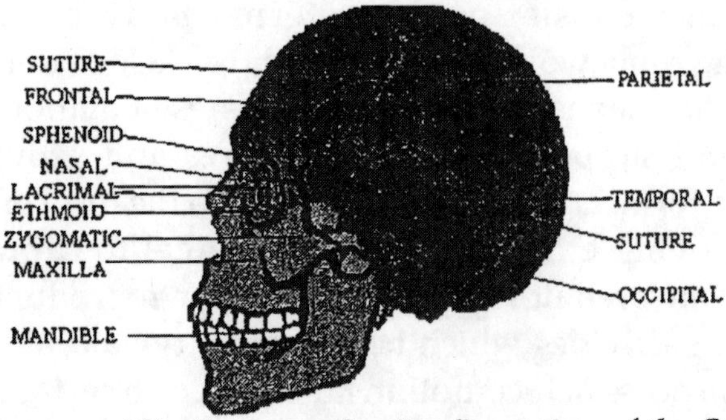

Fig. Sutures of Skull As per the Configuration of the Sutures, the Following Sub-types are Present

- Scarf joint. If the overlapping edges taper so that the two bones remain in line, it is called a scarf joint. This kind of a joint improves strength. (e.g. various skull bones).

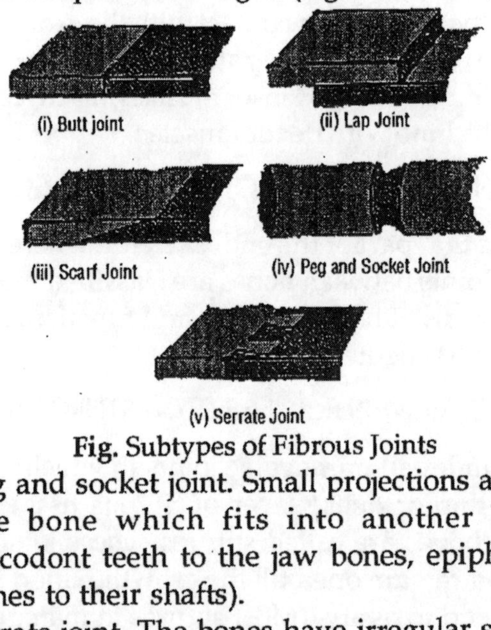

Fig. Subtypes of Fibrous Joints

- Peg and socket joint. Small projections are formed in one bone which fits into another bone. (e.g., thecodont teeth to the jaw bones, epiphysis of long bones to their shafts).
- Serrate joint. The bones have irregular suture, so the bones interlock repeatedly through the union. (e.g., roofing bones of the skull).

CARTILAGINOUS OR SLIGHTLY MOVABLE JOINTS (AMPHIARTHRIOSIS)

In this joint, the adjoining bony surfaces are connected by cartilage. It is a tough joint and allows some motion in response to compression, tension or twisting.

- The union between the bones of such a joint is called a symphysis instead of a suture (e.g., pelvic symphysis between pubis and pelvis, madibular symphysis, and vertebral centra, which provide for motions of the spine).

  The pubic symphysis is more flexible in women under the hormonal environment during pregnancy; this is needed for childbirth, to enable the birth canal to enlarge.

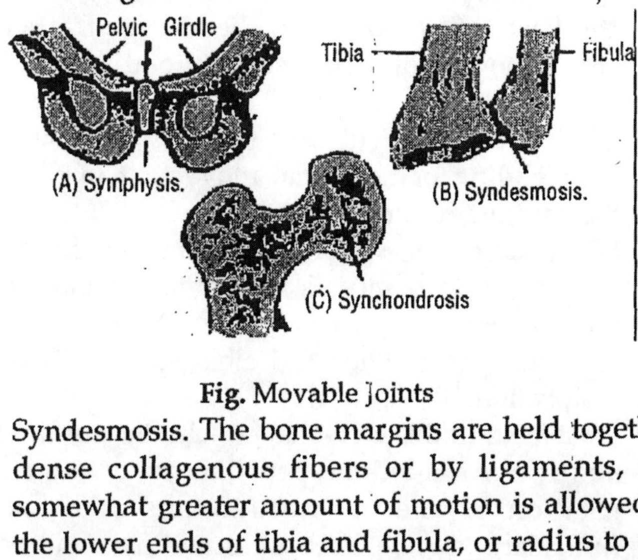

Fig. Movable Joints

- Syndesmosis. The bone margins are held together by dense collagenous fibers or by ligaments, and a somewhat greater amount of motion is allowed, (e.g. the lower ends of tibia and fibula, or radius to ulna).
- Synchondrosis is a cartilaginous joint in which the connecting cartilage is converted into bone. Thus, it is a temporary joint. (e.g., epiphysis and diaphysis of long bones). The resulting bony joint is called synostosis, which is an immovable joint.
- A permanent cartilaginous joint is present at the rib-sternum junction.

SINOVIAL OR MOVABLE JOINTS (DIARTHROSIS)

This is a joint in which a small space called a synovial cavity or joint cavity is present between the articulating surfaces of the bones. The bone ends are covered with a smooth hyaline cartilage which is lubricated by the synovial fluid. This is a clear, yellowish, viscous fluid which contains mucin. Its function is to nourish the hyaline cartilage and to lubricate the joints. The fluid prevents the friction between the solid surfaces in contact with each other. The synovial fluid is secreted by the synovial membrane which lines the joint capsules. The membrane contains fat cells which may form fat pads forming cushions to the joint. The synovial joints are freely movable and are made stronger by ligaments. The ligaments may be within or partly outside the joint cavity or may be inside the joint cavity (e.g., hip and knee of mammals).

On the basis of variations in the shape of articulating surfaces the synovial joints are sub divided into following types.

- *Gliding joint.* Articulating surfaces are usually flat, with gliding or non axial movement (e.g., between the carpal bones or vertebrae).
- *Hinge joint.* The spool-shaped surface fits into the concave surface and allow limited movement in one plane about a single axis (Uniaxial), like a hinge-door movement (e.g., knee and elbow joints). This resists dislocation.
- *Pivot joint.* In this arch-shaped joint the surface rotates about a rounded or peg-like pivot, and allows rotation of one bone around its own long axis (uniaxial), (e.g. between atlas and axis and between radius and ulna).
- *Condyloid joint.* An oval-shaped condyle fits into an elliptical cavity. The movement is in two planes (biaxial), at right angles to each other, back and forth and side to side.
- *Saddle joint.* A saddle-shaped bone fits into a socket that is concave-convex in opposite directions. It is a modified condyloid joint. Movement is biaxial (e.g. thumb, between metacarpal and carpal bones, heterocoelous cervical vertebrae of birds).
- *Ball-and-socket joint.* Ball-shaped head fits into a concave socket. It allows movement in many directions including rotation (triaxial) (e.g., hip joint

and shoulder joint). These joints are prone to sudden-strain and are easily separated or dislocated.

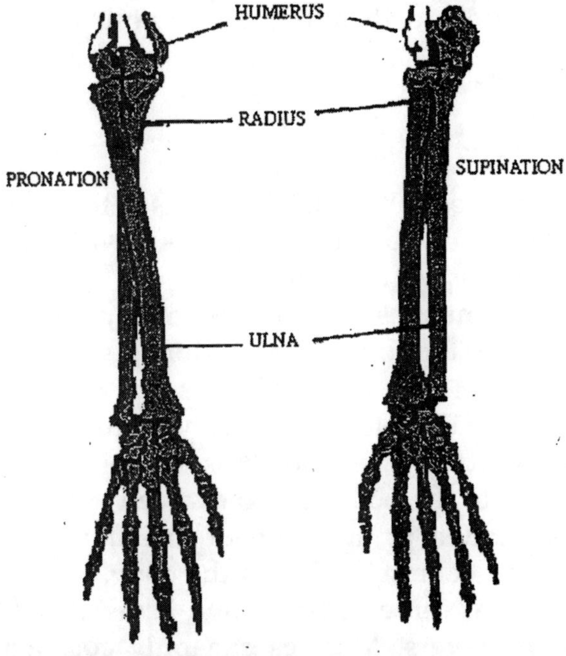

Fig. Sinovial Joints

- *Snap joints* are made stable by their ligaments which are lateral in position. Snap joints are found at the elbow and ankle of large mammals. Snap joints provide passive support in the standing position.
- *Plane joints* have more or less flat surfaces and allow various motion. These joints are found between the carpal bones of large mammals and between pre and post-zygapophysis.
- *Radial joints* are those in which one bone turns across another, as and when the radius of the forearm turns upon its own axis across the ulna in rotating the arm to turn the palm upwards (supination) or downwards (pronation).

Arthritis, the inflammation of a joint makes the joint painful and may immobilize movement at the joint. This results from a lack of synovial fluid, ossification of cartilage, deposition of uric acid crystals in the joint cavity or other changes due to age. A slipped disc is displacement of vertebrae from their normal positions. This results in pressing the nerves between the vertebrae, therefore it is painful.

MUSCLE MOVEMENTS

Skeletal or Voluntary Muscles

There are almost 700 skeletal muscles of the human body, governed by a few basic principles involving muscle movements or muscular activity.

- Skeletal muscles produce movements by pulling on bones or tendons. The tendon gives a very firm anchorage. The point where a muscle is connected to an immovable or relatively fixed bone is called the point of insertion. It is nearer to the central portion of the body. The movable bone is connected by a tendon at the point of insertion and is away from the central portion of the body.
- The bones serve as levers and joints act as fulcrums for the levers. Muscles can only contract a short distance, but since they are attached near a joint, the movement at the opposite end of a limb is greatly magnified. The biceps muscle of the arm may contract only 89 to 90 mm, but the hand will move about 60 cm.
- The muscles that move a part of the body usually do not lie over that part. For example, the biceps muscle that moves the lower arm does not lie over the lower arm (it is attached to it by tendons). But lies in the upper arm. Similarly, flexors of the wrist are located in the lower arm and not in the wrist.
- The skeletal or voluntary muscles act in pairs rather than singly. One of the muscles produces contraction while the other allows relaxation. Flexion results when contraction causes two bones to bend toward one another, while extention results from contraction of antagonistic muscles, resulting in an increase in angle between two bones. Such pairs of muscles are called

antagonistic. Often antagonistic muscles are in groups, e.g., both the brachialis and the biceps muscles flex the arm at the elbow and antagonize the triceps, but only when the palm is facing upwards.

Of such pairs or groups of antagonistic muscle, one is usually much stronger than the other. The biceps, which flex the arm are larger and more powerful than the triceps which extends it. The gastrocnemius and soleus muscles of the calf which extend the foot at the ankle are stronger than the peroneus muscles which flex it.

- When the body is at rest, the proportion of antagonistic skeletal muscles remain in a state of contraction, called muscle tone, which holds the body in rigid position. If the person becomes unconscious, or is asleep, muscle tone is lost as the muscles relax completely.
- For every action there is an opposing or restoring action (antagonists muscles). Rarely does one muscle contract alone, but is assisted by another. Synergists are muscles that supplement or assist one another, thereby reducing undesired action or unnecessary movement.

Involuntary Muscles

The involuntary or smooth muscles occur in the (intestines, gall bladder, urinary bladder ureter), arteries, iris, diaphragm of the eye etc.. They are abundant in sluggish invertebrates like molluscs. Structurally they are different from voluntary or skeletal muscles.

They are usually under automatic rather than voluntary control. In the alimentary canal etc. their distribution in circular and longitudinal muscles, allows an antagonistic action. Involuntary muscle produces relatively slow but prolonged contractions, or rhythmic contractions as in peristalsis.

Criteria for Naming Muscles

Muscles are often given names that describe their actions (levator, flexors, adductor extensors, abductors, depressors, sphincters, supinators, pronators.).

- By shapes (biceps, delltoid rhomboideus, trapezius).

- By positions (temporalis, pectoralis, gluteus, rectus femoris, tibialis anterior).
- By attachments, the origin being named before the insertion (geniohyoid, sternomastoid stylohyoideus.) While most muscles were first named in the human, the terms are not always appropriate when applied to homologs in other animals.

Chapter 6

Molecular Basis of Vision

COLOUR VISION AND PHOTOPIGMENTS

The eye was considered as an optical system which focuses images onto the retina. The retina was shown to act as a transducer converting the light to neural impulses. These, in turn, appear to be sorted and analysed both within the retina and within the brain to give rise finally to the sensation of vision.

Part of the visual sensation consists in recognizing different colours. The colour sensed is a function of the wavelength of the incident light; the complexity of this function is emphasized by the experiments. Nevertheless, any discrimination of different hues is due to the presence of receptors which selectively absorb light energy in certain wavelength regions. The photosensitive pigments are altered by the photons absorbed. These changes take place on a molecular level and are far too small to be revealed by any histological method. This particular aspect of vision is, therefore, in the realm of molecular biology.

All physiological or psychophysical experiments indicate that there must be at least three pigments within the retina of mammals. Whereas some experiments indicate that there may be many more than this, none indicate less than three. In spite of this, only two photosensitive pigments have been isolated from human retinas. One of these, rhodopsin, is found in the rods.

The molecular basis for mammalian vision is far from being completely understood; this contrasts sharply with the knowledge of the geometrical optics of the eye. Invertebrate vision has been even less well studied on a molecular level. Although it is evident that many insects possess highly developed colour senses which extend far into the ultraviolet, few of the photopigments responsible for insect vision have been isolated. As stated earlier, molecular biophysics is regarded by some biophysicists as the most fundamental part of biophysics. From this point of view, the most significant aspects of the biophysics of vision are the least well understood.

RHODAPSIN

It had been known for many years that pigments which might be associated with vision existed in the retina. In 1876, Boll named a pigment "Sehrot" which he saw in frog retinas. It was described as a brilliant red pigment. In 1878, Kuhne observed the bleaching of a pigment, rhodopsin, which he extracted with bile salts. Rhodopsin has also been called visual purple because of its characteristic colour. It is by far the most studied of the photosensitive pigments.

Today, rhodopsin has been purified and studied in many lobouratories. In order to extract it, fresh, dark-adapted retinas are mashed in a dim red light. The mashed retinas are then subjected to differential centrifugation until a fairly pure suspension of rods is obtained. The rods are hardened with alum, which makes most proteins insoluble. Then the hardened rods are extracted exhaustively with buffers to remove all water-soluble material, after which they are dried.

The rods are next extracted exhaustively with petroleum ether to remove all fat-soluble substances. This leaves insoluble particles which can be suspended only with suitable detergents. The particles containing the pigment rhodopsin are suspended in a 2 per cent solution of digitonin, or in bile salts. The entire purification must be carried out in a deep red light in order to have an appreciable yield of rhodopsin.

Is the rhodopsin one ends up with anything like the original? Tests show that it has almost the same absorption spectrum, although at physiological pH the peak is shifted several ml. This similarity is taken as evidence that the purified rhodopsin is essentially the same as the initial rhodopsin. If the extraction is performed in a bright light, no rhodopsin is present in the final suspension; only the protein, opsin, remains.

Rhodopsin is a protein complex with a molecular weight of approximately 40,000. It is a conjugated complex of a particular protein, opsin, with a much smaller hydrocarbon group, called neoretinene b or retinene. Retinene $_1$ is the aldehyde of vitamin A_1. Compounds with a ring structure of this type, plus a chain of about nine additional carbon atoms, are referred to as carotenoids because many of them are found in carrots. Vitamin A $_1$ is one of these carotenoids.

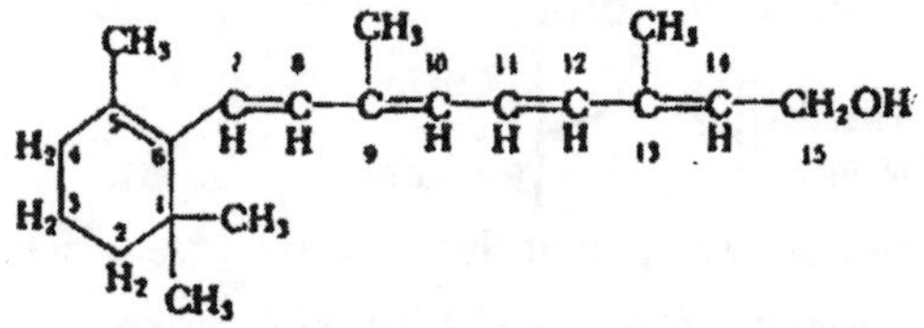

Fig. Structural Formula of vitamin A $_1$.

For every double bond along the vitamin A $_1$ chain, there are two spatial isomers. If one of the atoms on each carbon is hydrogen, one can represent the cis isomer as

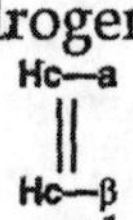

and similarly the trans isomer as

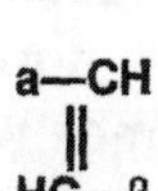

In many cases, it is possible to distinguish between the cis and trans compounds either by chemical or by X-ray techniques. In more complex cases, this distinction is difficult to demonstrate.

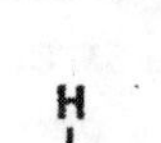

In the case of vitamin A_1 (or retinene), there are 2^4 or 16 different isomers which can be drawn with pencil and paper. Only one of these isomers is physiologically active.

Essentially, one may regard the visual process as the splitting of retinene from opsin and then the resynthesis of rhodopsin. Unfortunately, this is a gross oversimplification. For if the usual form of vitamin A is converted to retinene, it does not react with opsin at all. This is because the common vitamin A is the all-trans isomer. If retinene is split from rhodopsin, it is also the all-trans isomer and will not recombine. Chemical methods can show that the active retinene is a mono-cis compound but cannot distinguish between the four possible isomers having the cis configuration at the 7-8, 9-10, 11-12, and 13-14 positions, respectively.

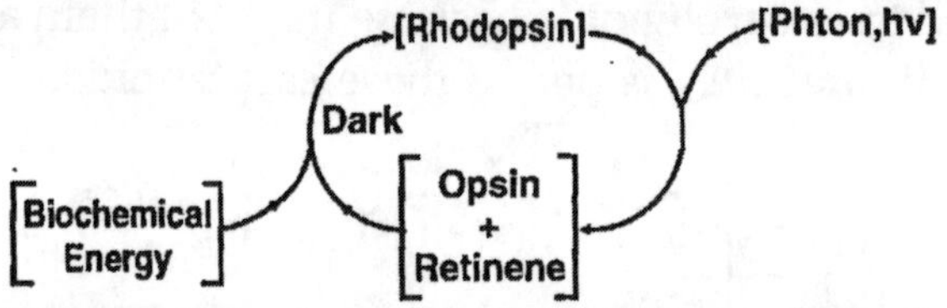

Fig. Simplified Version of the Visual Cycle of Rhodopsin.

On the basis of a study of the actual dimensional configurations, Pauling predicted that the only four stable isomers should be the all-trans, 9-cis, 13-cis, and 9-13 di-cis. Of these, only the 9-cis isomer reacts with opsin, and, although forming a photosensitive pigment, does not form rhodopsin. Thus, retinene cannot be one of the more probable forms.

The other mono-cis isomers, 7-8 and 11-12, are hindered forms. The interaction between different parts of the molecule twists the long side chain and the ring, distorting the normal planar form. Of these two hindered forms, the isomer with the cis bond at the 11-12 position is less hindered and so is more probable for the active form, which is called neoretinene b or retinene.

The plane projection of the 11-12 cis compound. This compound possesses a large steric hindrance between 13-CH_3 and the 10-H. The 7-8 cis compound possesses a still stronger hindrance between the 9-CH_3 and the 1- and 5-CH_3 groups. The conversion from the all-trans form of vitamin A aldehyde to neoretinene b occurs in the presence of iodine and light. It is believed to occur also within the retina where the conversion is enzymatically catalyzed.

There is another type of enzyme within the retina (and almost all other body tissues for that matter) which is known as alcohol dehydrogenase (abbreviated ADH). This enzyme interacts with the visual cycle because it acts as a catalyst maintaining an equilibrium ratio between the aldehyde retinene and the corresponding alcohol, vitamin A. This equilibrium is in the direction of much greater concentration of the alcohol. If, however, the aldehyde concentration is sufficiently low, the enzyme catalyzes the aldehyde production from the alcohol. To function, alcohol dehydrogenase needs a coenzyme, diphosphopyridine nucleotide (DPN^+) to accept the hydrogen atoms removed from the aldehyde. In the living retina, as in all other tissues, there is an abundant supply of DPN^+ and DPNH.

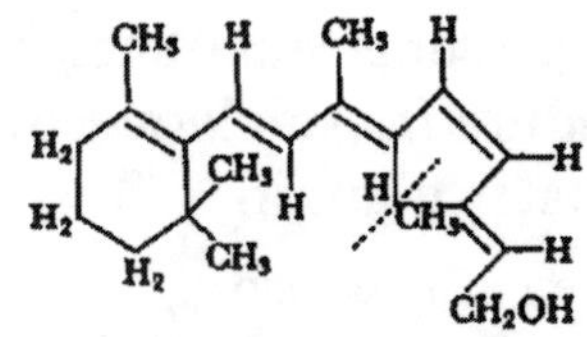

Fig. The 11-12 Mono-cis Isomer of Vitamin A_1.

The plane projection is distorted so that it does not show the steric hindrance. This is indicated by the broken line between the methyl group at 13 and the hydrogen attached to position 10. The vitamin A produced by the action of alcohol dehydrogenase and DPNH can pass through the rod membrane and into the blood stream. The vitamin A in the retina is in equilibrium with that in the blood under steady-state light conditions. The vitamin A in the blood stream is, in turn, maintained at a more or less constant concentration by the liver, which stores any excess of vitamin A.

The over-all process is then a complex cycle. In the dark, the cycle is stopped by all the opsin being bound in the form of rhodopsin. In the light, an equilibrium must be established with a steady-state concentration of rhodopsin. The concentration may be very close to zero, but a small part of the opsin should always exist in the form of rhodopsin.

From the point of view of the alcohol dehydrogenase, one may regard the opsin as a trapping reagent which effectively shifts the equilibrium so that the alcohol (vitamin A $_{1)}$ is converted to the aldehyde (retinene $_{1)}$. The reaction of retinene $_1$ with opsin is exothermic and goes spontaneously. From the point of view of the retinene, the opsin acts as an enzyme, converting the retinene with the help of the photon hv from the less probable cis form to the more probable all-trans form. One may describe the protein opsin as a photo-isomerase.

An additional feature of this reaction is that it tends to stabilize the reactants. Opsin and retinene$_1$ are both relatively unstable. For instance, by changing the pH of an opsin solution from a neutral 7.0 to either 5.0 or 8.0, opsin is 50 per cent irreversibly altered (denatured) in an hour. The cis retinene is relatively easy to "isomerize" to the all-trans form and is primarily converted to the alcohol (vitamin) form. In contrast to opsin and retinene, the compound rhodopsin presents remarkable stability as indicated by the extraction procedure. It is stable over the pH range 3.9 to 9.6. It is easy to imagine that in vitamin

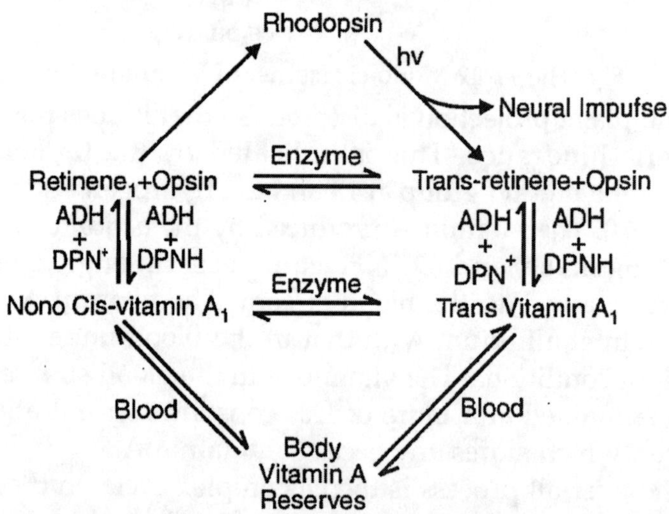

Fig. The Visual Cycle of Rhodopsin in the Retina.

A deficiency, the opsin might quickly degenerate. (Indeed, the rods do show very rapid degeneration.) On addition of vitamin A to the diet, the opsin formed would be stabilized as rhodopsin. Hence, the rods (and cones) could rebuild. This suggests that opsin is a type of adaptive enzyme. This mutual stabilizing may be typical of other adaptive enzymes.

Under certain conditions, the step from rhodopsin to trans retinene and opsin may be stopped at two intermediate points. If a dried film of rhodopsin is exposed at -70U+00B0C, a new compound, lumirhodopsin, is formed. On heating to 20U+00B0C, this spontaneously changes to metarhodopsin. This compound is stable when formed from squid rhodopsin. Metarhodopsin from vertebrate rhodopsin changes spontaneously in the presence of water to trans retinene and opsin.No evidence exists concerning the possible appearance of lumirhodopsin and metarhodopsin in the normal visual process. Perhaps partially bleached rhodopsin is converted to one of these intermediates under the conditions of colour vision and gives rise to an absorption curve different than the scotopic relative luminosity curve.

The nature of the bond between the carotenoid retinene and opsin has been investigated. Certain proteins are known to possess free —S—H groups called sulfhydryl groups. These are very reactive with a variety of compounds but are inhibited by the compound para chloromercuribenzoate (PCMB). The reaction of opsin and retinene is also inhibited by PCMB; quantitative studies indicate that the reaction may involve two sulfhydryl groups per retinene molecule. This, however, may be an oversimplification. During the reaction, the pH changes in a manner which corresponds to blocking or tying up a base such as histidine. The detailed reaction of opsin and retinene is not known.

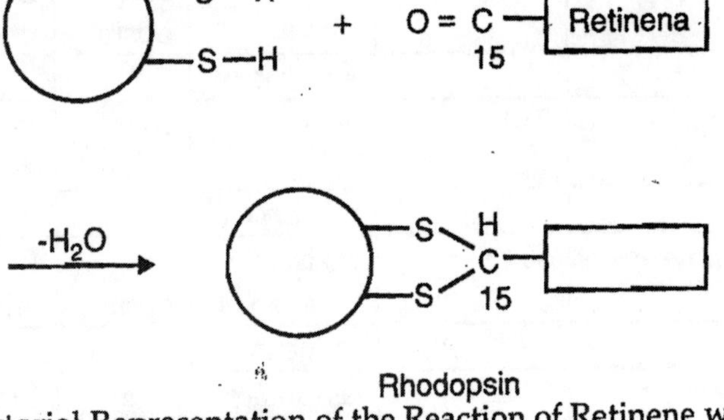

Rhodopsin

Fig. Pictorial Representation of the Reaction of Retinene with the Postulated Sulfhydryl Groups of Opsin.

OTHER PHOTOPIGMENTS

The cones of the human eye have yielded another pigment similar to rhodopsin but having a different absorption maximum. This pigment, called iodopsin, is of uncertain physiological action because its spectrum does not correspond to any observed physiologically. Talbot avoided this issue by adding a contribution labeled "dz" from the rods and the hypothetical 8 cones. Because the latter are only postulated, one certainly can account for any physiological spectrum, but this theory lacks a conviction similar to that presented by the similarity of the scotopic luminosity curve and the rhodopsin absorption curve.

Pigment	Absorption maximum mμ	Carotenoid	Protein	Source
Rhodopsin	500	retinence$_1$	rod opsin	Land vertebrates, marine fishes
Iodopsin	562	retinence$_1$	cone opsin	Birds, turtles, mammals
Porphyropsin	522	retinence$_2$	rod opsin	Fresh water fish
Cyanopsin	620	retinence$_2$	cone opsin	Turtle, tench
Isorhodopsin	487	9-cis retinence$_1$	rod opsin	In vitro only
Isoiodopsin	510	9-cis retinence$_1$	cone opsin	In vitro only
Isoporphyropsin	507	cis-isomer retinence$_2$	rod opsin	In vitro only
Isocyanopsin	575	cis-isomer retinence$_2$	cone opsin	In vitro only

An examination of iodopsin (from cones) shows that it is very similar to rhodopsin. Iodopsin is a conjugated compound consisting of retinene1 and another protein, cone opsin or photopsin. (The rod protein is referred to in various places as rod opsin and scotopsin.) The rod opsin and the cone opsin extracted from a wide variety of animals, including all the vertebrates, are essentially similar. Some animals have a different aldehyde called retinene instead of retinene. Retinene $_2$ is the aldehyde of vitamin A_2, which differs from A_1 in having an extra double bond in the ring between 3-C and 4-C.

Retinene is also a mono-cis compound, but the location of the cis bond has not been determined. A list of some of the known pigments of this general type is given in the accompanying table. Birds and reptiles have mostly cones in their retinas and probably lack rods. These cones contain bright coloured oil globules between the inner and outer cone segments. These oils are all carotenoids. Wald extracted 1,600 bird eyes to isolate the following oils

Carotenoid	Absorption maximum, mμ	Color
Astacin	497	purplish red
Xanthophyll	463	golden yellow
Galloxanthine	450	greenish yellow

Colour vision is clearly possible with just these filters and iodopsin.

THE ORIGIN OF THE NEURAL SPIKE

Many experiments indicate that one photon dissociates one retinene $_1$ group from one rod opsin molecule. It is completely unknown how one or two such dissociations can give rise to a neural spike. This is one of the most fundamental questions which may be asked about the molecular changes in vision. One suggestion is that somehow the retinene alters the permeability of the rod when it is dissociated from the opsin.

The rhodopsin in a fixed rod has been shown by electron microscopy to be arranged in a regular array of disks which essentially fill the rod. The arrangement of rhodopsin molecules within the disks is not known. The disks are about

250 mì diameter and 50 mì thick. It is possible that this array of disks acts somehow as a semiconductor (or even transistor) whose conduction depends on the number of impurity centres (dissociated retinene molecules). But this cannot be proved. In common with hearing, olfaction, and taste, it is impossible to describe the method by which the neural impulse is started.

Chapter 7

Respiration and Blood Circulation

All organisms, without exception, require energy for sustaining life activities. Respiration is an energy releasing process. It is defined as biological oxidation in which stored food is broken down step by step to release metabolically usable energy, giving out CO_2 and water vapour as by-products. It goes on in all living cells at every moment of their existence as living entities. Physiologists distinguish respiration into external and internal respiration.

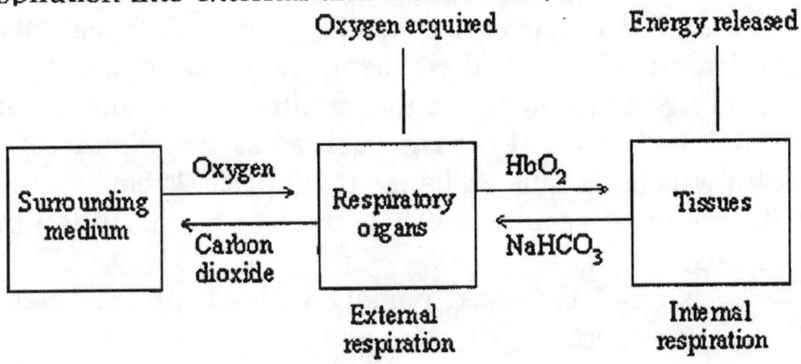

Fig. Respiration

- External respiration or breathing: It involves the exchange of gases between surrounding medium and blood circulating in the respiratory organs. It is a physical process in which free oxygen is acquired by the animal. It neither involves oxidation nor release of energy.
- Internal respiration or cellular respiration: It takes

place in the cells during which the food is oxidized and energy is released. Biological oxidation is not the process of direct oxidation, but a complex oxidation-reduction process is brought about by a series of respiratory enzymes and coenzymes present in the mitochondria. The overall reaction is as follows:

GASEOUS EXCHANGE AND TRANSPORT

GASEOUS EXCHANGE

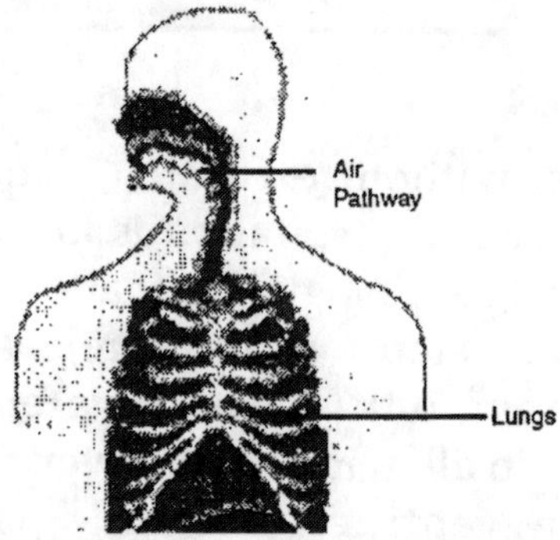

Fig. Respiratory Organs of Man

In unicellular animals like Ameba, gaseous exchange takes place directly through the plasma membrane and in its surrounding medium, but higher multi-cellular animals are provided with special organs such as skin, gills or lungs through which gaseous exchange takes place. Whatever may be the organ of respiration it is subject to the following conditions:

- It should be naked namely without an exoskeleton of any sort.
- It should have great absorptive surface exposed to the surroundings to absorb enough oxygen to fulfill demands of the body cells.
- It should be covered with a thin membrane permeable to gaseous exchange.
- It should be richly supplied with blood capillaries.

HUMAN RESPIRATORY SYSTEM

The human respiratory system consists of two parts:

- Respiratory pathway.
- Respiratory organs (lungs).

Respiratory pathway It includes nasal cavities pharynx, larynx and trachea.

- Nasal cavities are a pair of cavities in the nose, separated by a vertical cartilaginous septum and lined by ciliated mucous membrane. While passing through these cavities air is filtered, moistened and warmed. The nasal cavities open posteriorly into the pharynx.
- Pharynx serves as a common passage for the transport of air to the larynx, and food from the mouth to the esophagus.
- Larynx (commonly known as "Adam's apple") is situated in the neck just at the start of the trachea. It is guarded against the entry of food particles by a flap-like valve called the epiglottis. The larynx contains the vocal cords hence it also serves as the organ of phonation (i.e. producing sound).
- Trachea or windpipe is a tube about 4"long running from the larynx to the lungs. It is supported by C-shaped cartilages which keep the trachea distended. It is internally lined by cilia beating upwards, driving out dust particles. On entering the thorax, the trachea divides into two branches called bronchi, each entering the lung on its side.

Respiratory Organs (lungs)

The lungs are a pair of essential respiratory organs situated in the chest cavity one on each side of the heart and surrounded by a double membrane, the pleura. They are protected laterally by ribs, interiorly by the sternum and posteriorly by the vertebral column.

They are separated from the abdomen by the transverse muscular partition called thediaphragm. The lungs are smooth, spongy and pink in colour. The right lung is divided into three lobes, while the left lung is divided into two lobes: it is smaller so as to accommodate the heart in between.

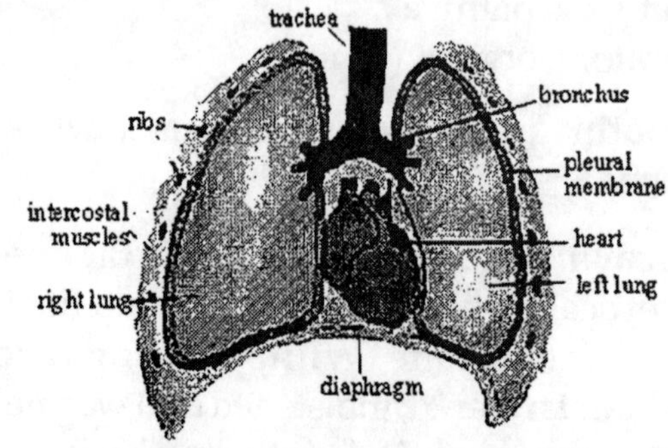

Fig. Situation of the Lungs

The lung is mainly formed by bronchioles and alveoli. Each bronchus after entering the lung divides into smaller and smaller bronchioles each about a millimeter or less in diameter, like a tree branching into thinner branches forming slender twigs. The ultimate branches of the bronchi called terminal bronchioles leading to respiratory bronchioles, end in, clusters of alveolar sacs each bearing a number of alveoli looking like a bunch of grape-like structures.

The branches of the pulmonary artery and pulmonary vein, which divide and re-divide along with the branches of the bronchi, ultimately form a capillary network around the alveoli. Thus the lungs are essentially made up of a complex branching system of bronchi ending in alveoli, bound together by connective tissue and containing a network of blood capillaries. The alveoli are lined throughout by flat epithelial cells to facilitate exchange of gases. They are surrounded by a rich and fine network of blood capillaries.

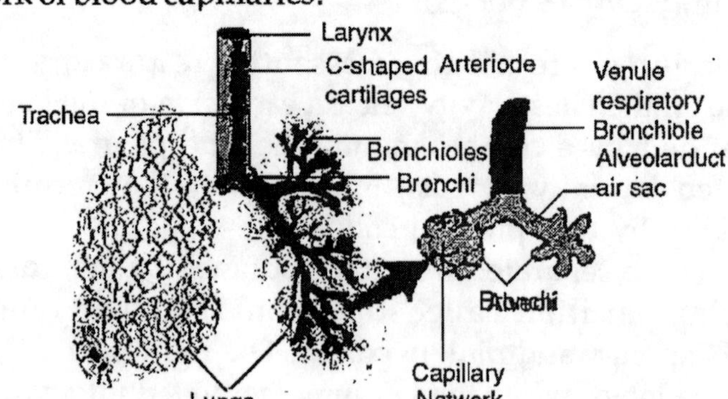

Fig. (A) Structure of lungs (B) Part of the lung enlarged

There are about 750 millions of these alveoli, the combined area of which is about 100 square meters, i.e., about a hundred times the area of skin on the human body. Thus a vast surface is exposed for exchange of gases in the lungs. A transverse section of a lung shows a number of bunches of grape-like structures called alveoli lined by flat epithelial cells showing connective tissues and blood capillaries in between. It also shows a section of terminal bronchioles lined by smooth muscle fibers. Respiratory bronchioles and blood capillaries may also be visible.

MECHANISM OF BREATHING

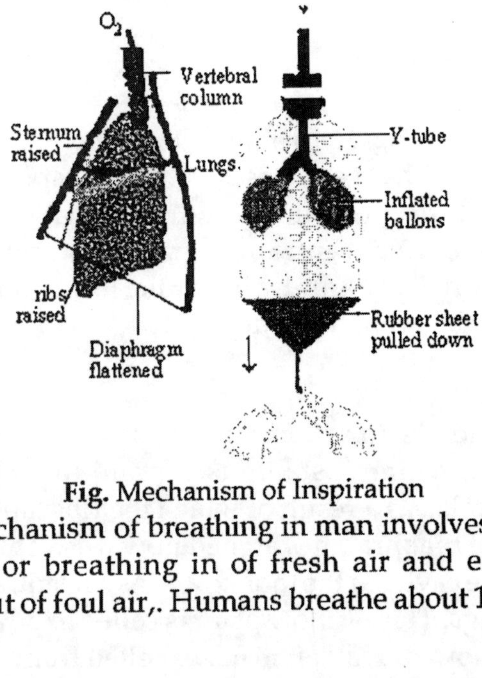

Fig. Mechanism of Inspiration

The mechanism of breathing in man involves two events: inspiration or breathing in of fresh air and expiration or breathing out of foul air,. Humans breathe about 16 to 18 times per minute.

Inspiration

Inspiration in man is an active process. During inspiration the intercostal (in between the ribs) muscles contract so that the ribs swing outwards and upwards and increase the diameter of the thorax. At the same time the dome-shaped diaphragm is flattened due to the contraction of its muscles. The combined effect is to increase the capacity of the thorax which creates a partial vacuum, as a result of which the air is drawn into the lungs.

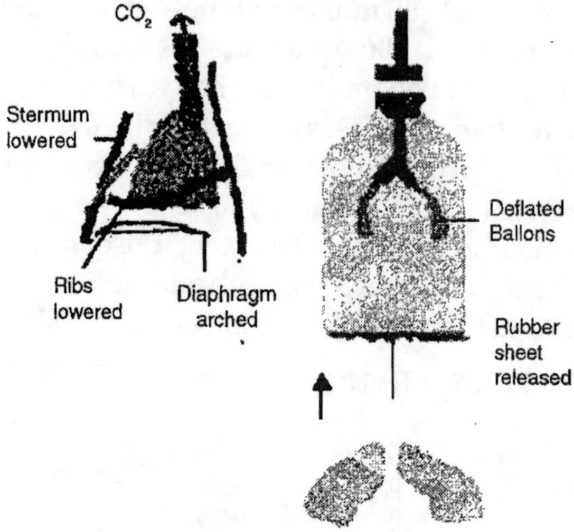

Fig. Mechanism of Expiration

In lungs, exchange of gases between inspired air and the bloodstream takes place through the internal lining of alveoli due to different concentration gradients and the enzyme carbonic anhydrase, and air in the lungs soon becomes saturated with carbon dioxide.

Expiration

Expiration is an automatic or passive process. During expiration, the intercostal muscles and the muscles of the diaphragm relax as a result of which the ribs fall, the sternum sinks and the diaphragm is arched upwards. This reduces the volume of the cavity of the thorax, and as a result, air is expelled from the lungs. This outflow of air is called expiration. During expiration, however, all air is not expelled from the lungs, but some residual air is always left in the alveoli.

Amount of air Breathed in and Breathed out

- Tidal air: It is the amount of air breathed in and breathed out during restful breathing. It is about 500 ml.
- Reserve air: It is the amount of air which remains in air passages after expiration. It is about 150 ml.
- Residual air: It is the amount of air which remains in the alveoli after forced maximal expiration.

EXCHANGE OF GASES

The physiology of respiration involves exchange of gases in lungs and tissues and their transport from lungs to the tissues and vice versa. Exchange of gases takes place by diffusion mainly based on concentration gradient.

Transport of Oxygen

Oxygen is transported from the respiratory organs to the tissues mainly by the respiratory pigment, hemoglobin (Hb).

Oxygen which is acquired during respiration diffuses from the alveoli of the lungs into blood due to its high concentration in alveoli. Hemoglobin of the red blood corpuscles has great affinity for oxygen. In blood, oxygen combines with hemoglobin to form an unstable reversible compound called oxyhemoglobin.

$$\underset{\text{Homoglobin}}{Hb} + \underset{\text{Oxygen}}{O_2} \rightleftharpoons \underset{\text{Oxyhemogiobin}}{HbO_2}$$

When blood reaches tissues, HbO_2 dissociates into hemoglobin and oxygen due to lower oxygen levels in the tissues, and the oxygen released diffuses into the cells. According to Bohr effect, the variations in the partial pressure of CO_2 in the tissues also helps in this dissociation.

Transport of CO_2

CO2 is transported by blood from the tissues to the lungs. It is carried in the following forms:

- Bicarbonate of sodium in plasma and potassium in red blood cells.
- Carbaminohmoglobin in red blood cells.
- Simple physical solution in plasma.
- About 85% of CO_2 is transported in the form of bicarbonate.

From the tissues CO_2 diffuses into blood, and combines with water to form carbonic acid.

$$\underset{\text{Carbondioxide Water}}{Co_2} \rightleftharpoons \underset{\text{Carbonic acid}}{H_2CO_3}$$

Since red blood cells contain the enzyme carbonic anhydrase which accelerates the reaction, most of CO_2 is collected in the RBCs than in plasma. Carbonic acid further dissociates into bicarbonate (HCO_3) and hydrogen ions (H^+).

$$H_2CO_3 \rightleftharpoons H^+ + HCO_3$$

Most of the HCO_3 diffuses into plasma and reacts with sodium ions (Na^+) to form sodium bicarbonate and is carried to the lungs.

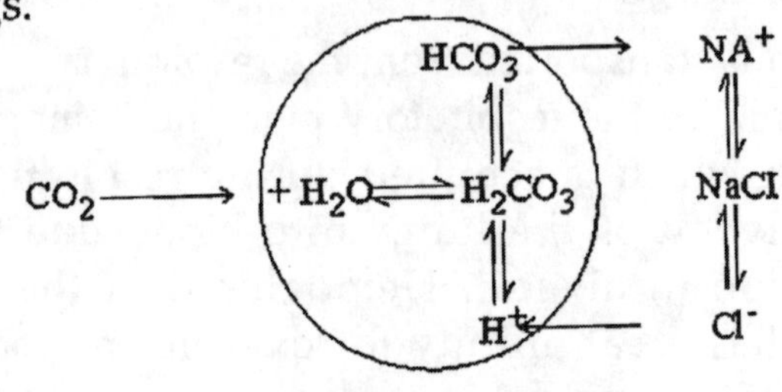

RED CORPUSCLE

Fig. Chloride Shift

$$NaCl \longrightarrow Na^+ + Cl$$

$$Na^+ + HCO_3 \rightleftharpoons NaHCO_3$$

As bicarbonate ions (HCO_3) diffuse into plasma from RBC's, Hydrogen ions (H^+) concentration in RBCs increases. For maintaining ionic balance for each bicarbonate (HCO_3) one chloride ion (Cl) from plasma enters the RBC. This is called "chloride shift". H^+ ions in RBC's also bring about dissociation of oxyhemoglobin (HbO_2) into Hb and O_2. The oxygen released diffuses into the tissues while hemoglobin buffers hydrogen to form very dilute hemoglobinic acid.

About 10% of CO_2 combines with an amino group (NH_2) on hemoglobin to form carbaminohemoglobin. This enters the lungs and dissociates into hemoglobin and CO_2, the latter of which then diffuses into the alveoli.

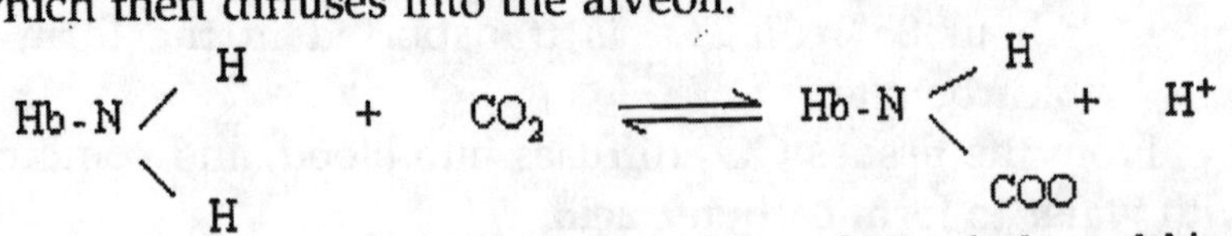

Hemoglobin Carbondioxide Carbaminohemoglobin

Nearly 5% of CO_2 never enters RBC but dissolves in plasma, where it combines with plasma protein to form proteinic acid which is carried to the lings where it dissociates into protein and CO_2. CO_2 further enters alveoli of the lungs.

In the lungs CO_2 tension is less than in the blood stream, hence the reversible reactions take place. Bicarbonate enter RBC and because of the enzyme carbonic anhydrase, it dissociates into H_2O and CO_2. This dissociation is facilitated by acidic oxyhemoglobin present at the respiratory surface. As a result CO_2 diffuses into alveoli and is expelled during respiration.

BLOOD CIRCULATION

The transport system of an animal moves substances to where they are needed in the body. Even the smallest animal must have the means of transporting substances around its body. Oxygen and food molecules must move to all the cells, and the waste products must be removed from the cells and expelled into the environment. A variety of fluid systems, called vascular systems, help such transport in most members of the animal kingdom. A circulatory system is a vascular system (i.e. with tubes or vessels) in which the transport fluid moves rhythmically in a particular direction, since it is propelled by a muscular pumping structure.

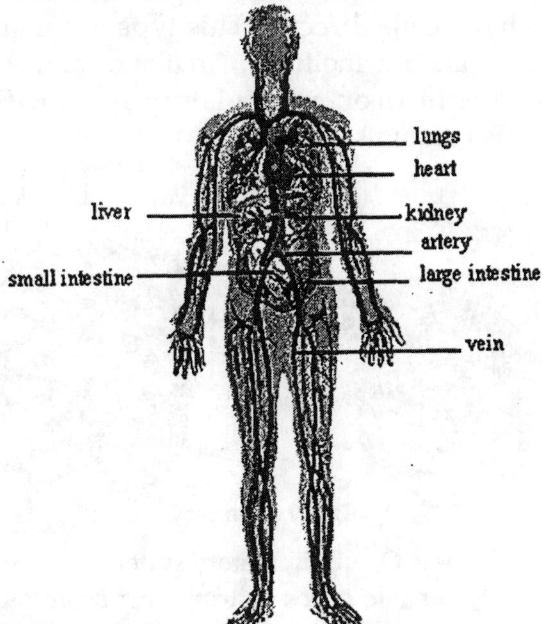

Fig. Vascular Systems

Unicellular organisms (Protozoa) have no specialized

system for circulation of fluid. Sponges and coelenterates also lack a specialized circulatory system. Part of the evolution of multicellular animals was the development of the body fluids (tissue fluids, blood and lymph). These fluids provide all body cells a stable and relatively non-fluctuating environment, and fill in the space between cells and cell layers.

The body fluids are divided into

- Intracellular (inside the cells)
- Extracellular, (blood plasma and interstitial fluid).

The blood plasma is present in blood vessels and the interstitial fluid in the spaces around the cells. Nutrients and gases passing between blood vessels and cells have to cross this fluid.

VASCULAR SYSTEM

- In an open (vascular) circulatory system the blood is not completely enclosed within vessels and it flows within blood vessels for only a limited part of its circuit. The blood vessels open into the open fluid spaces (sinuses or hemocoel) so that the circulating fluid (hemolymph) can reach the cells directly. This type occurs in insects, crustaceans and molluscs. Circulation and O_2 transport is slower in an open circulatory system. (Smaller or sluggish animals).

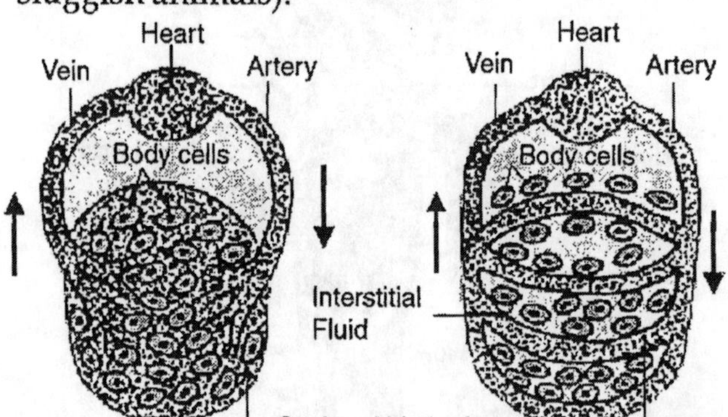

Fig. Open Circulatory system

There are two basic types of transport systems:

- In a closed circulatory system the blood remains within a completely closed (unbroken) system of vessels, and never comes in direct contact with the cells. This occurs in molluscs (squids), annelids (earthworms), echinoderms, and vertebrates (animals and humans). The advantages of a closed circulatory system are: faster transport of oxygen; greater efficiency of blood flow; economy of blood volume and maintenance of sufficient blood pressure for a large body.

A circulatory system, whether open or closed, has several functions such as transportation of dissolved nutrients, hormones, gases, antibodies, control of fluid volume, pH, and regulation of body temperature.

HEART

Position and External Features of Heart

The human heart is a muscular, four chambered organ about the size of a clenched fist. It is located in the thorax between the lungs (mediastenum) and above the central depression of the diaphragm. It is somewhat cone-shaped with the broader end (base) upwards, and the blunt end (apex) lying on the diaphragm. It is slightly oriented towards the left. The heart weighs about 300 grams in males, and is usually slightly smaller in females.

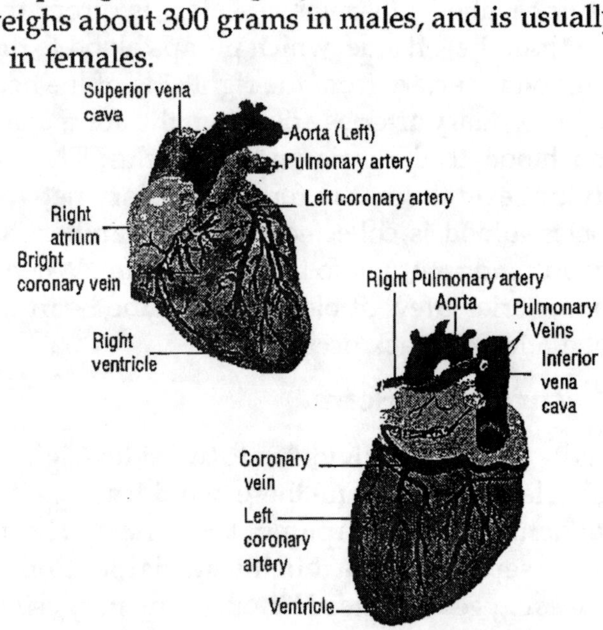

Fig. Ventral or Sternocostal View of the Heart.
Dorsal or Posterior View of the Heart

The heart is enclosed in a tough, double walled membrane

called the pericardium. The outer tough layer of pericardium is fibrous and forms a protective sheath. It is attached by ligaments to the breastbone and spine, anchoring the heart in position. Inside the fibrous pericardium is a thin but tough double membrane, the parietal pericardium. The pericardial fluid between the two membranes prevents friction, protects it from external shocks and over-distention of the heart with blood in extreme conditions.

The heart consist of four chambers, two atria (auricles) and two ventricles. The atria are thin walled and receive blood from four pulmonary veins (auricles) and two ventricles. The atria are thin-walled and receive blood from four pulmonary veins and two vena cavae. The main tasks of the atria are to prime (fill) the ventricles between "pumps", and to help monitor blood pressure in the vascular system. The ventricles are thick walled and muscular. The walls of the left ventricle are at least three times thicker than those of the right, as it pumps blood to the whole body through the aorta.

Two thick-walled muscular vessels arise from the heart – the aorta from the left side, which pumps blood to the body, and the pulmonary artery from the right side of the heart. The left and right coronary arteries arise from the aorta and supply oxygenated blood to the atria and ventricles. The coronary arteries divide several times to form a capillary network over heart muscles. Blood is collected from the capillaries by the coronary veins and returned to the right atrium. When defects in coronary arteries prevent blood flow to the heart muscles, that is when a heart attack occurs.

Internal Structure of the Heart

Internally, the heart is divided into two sides (right and left) by two muscular partitions, and these two sides work in unison but independently as a double pump. The auricles. The two atria or auricles are separated by a thin inter atrial septum. The left atrium receives oxygenated blood by four pulmonary veins, which opens into its upper part. Their openings are without valves. The left atrium opens into the left ventricle by atrio-ventricular orifice (opening), guarded by the bicuspid or mitral valve.

The right atrium receives deoxygenated blood from the body through the superior vena cava (opens into the upper part), and inferior vena cava, which opens into the lower part of the atrium. Its opening is guarded by a rudimentary valve. A small opening of coronary sinus (vein), guarded by a valve, is present between the orifice of the inferior vena cava and atri-ventricular opening. The fossa ovalis is a small oval depression above and to the left of the opening of the inferior vena cava. The right atrium opens into the right ventricle, and this opening is guarded by tricuspid valves to prevent back-flow of the blood from the ventricle to the atrium.

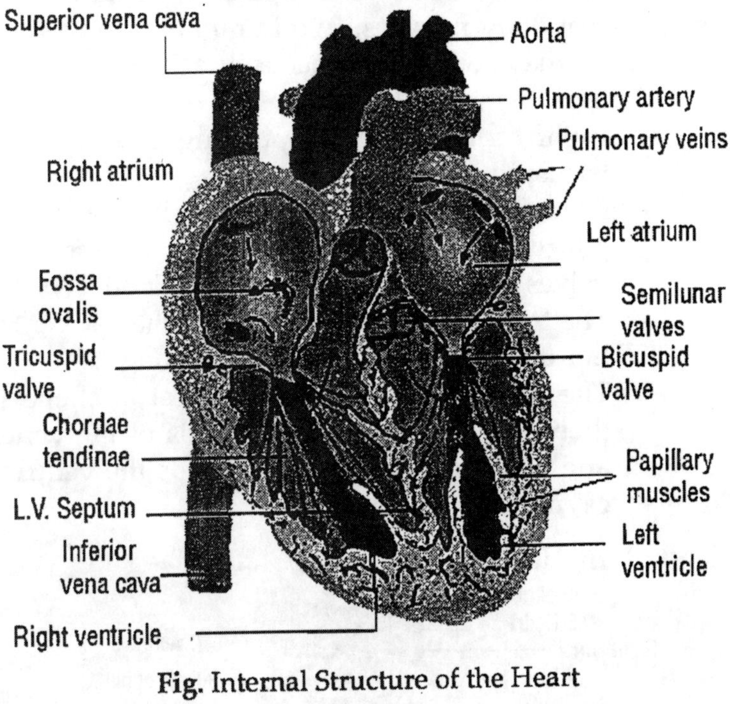

Fig. Internal Structure of the Heart

The Ventricles.

The right and left ventricles are separated by inter ventricular septum. The septum slopes obliquely with convexity towards the right ventricle. The left ventricle is longer and more conical than the right, and forms the apex of the heart. The left ventricle shows the following important features:

- The left atrio-ventricular opening guarded by a mitral or bicuspid valve.
- A circular opening of the aorta guarded by aortic or semilunar valves. The valves allow the blood to enter the aorta from left ventricle during ventricular contraction and prevents it from flowing back into the ventricle during relaxation. The aortic valve has 3 cups-two posterior (right and left) and one anterior.
- The trabeculae carneae (chordae tendinae), are attached to the margins of the bicuspid valve and prevent them from everting into the atrium.
- The papillary muscles, two in number, to which one end of the chordae tendinae is attached.

  The right ventricle has:
  - Right atrio-ventricular opening guarded by the tricuspid valve.
  - A rounded opening of the pulmonary artery, guarded by semilunar valves. The semilunar valves prevent backflow of the blood.
  - The trabeculae carneae are as in the left ventricle, but they are not as stout and strong.
  - The papillary muscles are conical in shape with their bases attached to the walls of the ventricle and their apices directed towards the ventricular cavity.

Working of the Heart

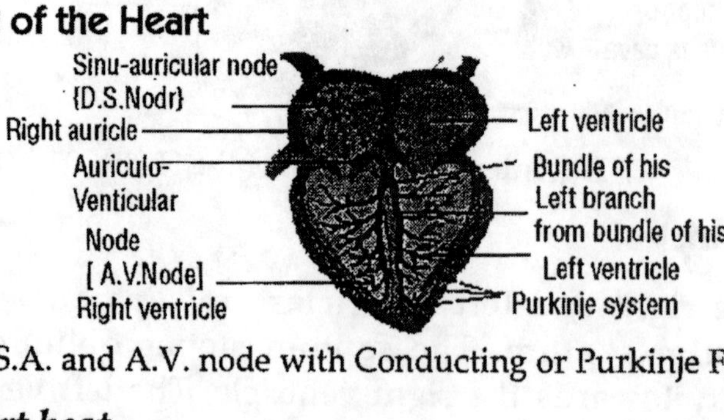

Fig. S.A. and A.V. node with Conducting or Purkinje Fibers

The heart beat

The heart is a remarkable organ. The heart beats continuously throughout an animal's life. The heart contracts rhythmically, pumping a certain volume of blood from each ventricle to different parts of the body. This volume is called cardiac output. The contraction of the heart is called systole and the relaxation, diastole. The contraction and relaxation of the heart forms a heart beat. The human heart beat can be heard as "lub-dub" sound (heart sound) with a stethoscope. When the body is at rest, the heart beats approximately 70 times per minute, and each cardiac cycle lasts for about 0.8 seconds.

Functioning

- The cardiac cycle (heart cycle) begins with atrial contraction (atrial systole), which is initiated by electrical stimulus coming from the sino-atrial (S.A) node, or "pacemaker". The stimulus is transmitted through the atrio-ventricular (A.V.) node to bundle of His, which spreads to form Purkinje fibers on the ventricle. During atrial contraction, the blood from the atria is forced into the respective ventricles. The bicuspid and tricuspid valves prevent back flow of the blood.
- The ventricles now undergo powerful contraction (ventricular systole), during which bicuspid and tricuspid valves remain closed. When the ventricular blood pressure increases (systolic pressure) to more than that in the aorta and pulmonary arteries, the blood enters these arteries through the semilunar valves.
- When the blood enters the arteries, the ventricles relax (ventricular diastole) and the pressure (diastolic pressure) in them falls. Simultaneously the atria have relaxed (atrial diastole) and the atria fill with blood again. The cardiac cycle continues in this way.

Heart sounds

During ventricular systole, simultaneous closure of atrio-ventricular valves makes the first sound –"lub". The second sound – "dub"- which is higher-pitched, shorter and sharper, is caused by the simultaneous closure of aortic and pulmonary valves. Extra heart sounds, called murmurs or a "hiss" sound, occurs when blood leaks from the artery back into the ventricle, usually because one of the valves is defective. (In that case the rhythm would be: Lub-hiss-dub or lub-dub-hiss.)

ARTERIAL BLOOD PRESSURE

Blood pressure may be defined as the pressure exerted by the blood against the vessel walls which contain it. This pressure is greatest in the arteries during ventricular contraction (systole), and is called systolic pressure. During ventricular relaxation (diastole), blood pressure falls, reaching a minimum pressure just prior to the next systole. The minimum pressure is referred to as diastolic pressure. Average arterial pressure during systole is about 120 mm Hg in an adult, while the diastolic pressure is about 80 mm Hg. This is normally expressed as blood pressure of 120/80, the upper number indicating systolic pressure and the lower number indicating the diastolic pressure.

Blood pressure is measured by an apparatus called the sphygmomanometer, which measures blood pressure in an artery. When the blood pressure is taken, the cuff is wrapped around the arm just above the elbow over the brachial artery, and air is pumped into the bag. A stethoscope is placed over the artery so that the pulse can be read. Air is pumped untill flow stops in that artery and the sounds cease.

Now the bag is slowly deflated until blood starts to flow and the pulse (sound) can just be heard. The manometer reading at this time indicates systolic pressure. Deflation of the bag is continued and a reading just before the last pulse sound is taken. This indicates diastolic pressure i.e. lowest arterial pressure caused by the ventricular diastole

Extreme deviations from the normal blood pressure (120/80) are indications of a heart malfunction, unusual blood volume in the system, arterial inelasticity (arteriosclerosis), kidney disease, etc. A decrease in the blood pressure below the normal level is referred to as hypotension or low blood pressure. This may happen due to low blood volume or sometimes due to a defect of the heart. A persistant increase in blood pressure above the normal level (say 160/95), is referred to as hypertension or high blood pressure. People with high blood pressure are susceptible to stroke, heart diseases, kidney failure, headaches, etc.

BLOOD

Composition of Blood

Blood is a viscous fluid of bright red colour, heavier than water and slightly alkaline. It consists of two components. The liquid part is called plasma, and the large number of cells is called blood corpuscles suspended in the plasma.

Plasma

It is a straw-coloured fluid consisting of the following chemicalonents:

- 90% of water.
- 7% of plasma proteins like serum albumin, serum globulin and fibrinogen.
- 3% of remaining substances, such as:
  - Nutrients like glucose, amino acids and fatty acids.
  - Inorganic salts like bicarbonates of sodium, potassium, calcium and magnesium.
  - Organic compounds like enzymes, hormones, antibodies, antitoxins and heparin, etc.
  - Gases like oxygen, carbon dioxide and nitrogen.
  - Waste materials like urea, uric acid, etc.

The composition of plasma is altered in tissues and controlled by organs like the liver and the kidneys.

Functions of Plasma

Plasma is the chief transporting medium supplying water, Oxygen and metabolites to all cells and also collects nitrogenous waste products and Carbon Dioxide from them.

Blood Corpuscles

These are blood cells suspended in plasma and are of three types, namely, erythrocytes, leucocytes and blood platelets.

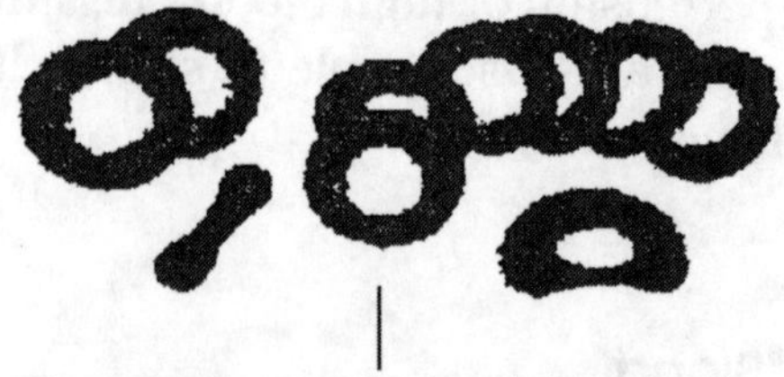

Fig: Human Red Blood Corpuscles

Red blood cells (corpuscles) or Erythrocytes

Red blood cells or erythrocytes are red in colour and are numerous, about five million per cubic millimeter; hence they give the red colour to blood. They are circular biconcave, anucleate (without a nucleus) and immobile. Being flexible, they get twisted and can be squeezed through blood capillaries smaller than their diameter. Each red blood cell shows the central spongy mass called stroma containing a red coloured substance called hemoglobin which is the respiratory pigment. Hemoglobin readily combines with oxygen to form an unstable compound known as oxyhemoglobin.

$$\text{Hemoglobin} + \text{Oxygen} \rightleftharpoons \text{Oxyhemoglobin}$$

Hence red blood cells are called oxygen carriers. The red blood cells of human beings are short-lived, having about 100 to 120 days of normal life span, probably because of the absence of a nucleus. They are produced in the red bone marrow and when dead they are constantly disintegrated in the liver and spleen.

Hemoglobin contains iron atoms involved in binding oxygen; thus liver is a good dietary source of iron.

Functions of Red Blood Cells (RBC)

- To carry oxygen in the form of oxyhemoglobin from the respiratory organs to all cells of the body.
- To carry carbon dioxide in the form of potassium bicarbonate from the tissue cells to the respiratory organs.

Leucocytes or White Blood Corpuscles

White blood corpuscles are less numerous, about seven

thousand per cubic millimeter of blood, i.e. the ratio of RBCs to WBC (White Blood Cells) is about 600:1. They are colourless, nucleated and migrate anywhere in the body. Leucocytes are produced in spleen or yellow bone marrow and when worn out, they are degraded in the blood stream and engulfed by other phagocytes.

Some leucocytes are also destroyed by microorganisms at the site of invasion (infection). These destroyed leucocytes, along with damaged tissue cells with some plasma, form yellowish liquid called pus, causing inflammation or abscess. During infection the number of leucocytes in blood increases. A chronic abnormal increase in leucocytes can lead to leukemia or blood cancer inhibiting the normal production of RBCs.

Functions of leucocytes

- They are mainly phagocytic in function ingesting bacteria, cell debris and other injurious materials occurring in blood.
- They produce antibodies and antitoxins which are specific in action.

Leucocytes are mainly of the following two types:

Granulocytes

They are distinctly amoeboid, containing granular cytoplasm and a lobed nucleus in each. According to their staining properties, they are classified as follows:

- Eosinophils: These are the granulocytes the granules of which are large and can be stained with acidic dyes. The nucleus is bilobed. They are about 5% of the total leucocytes. The percentage of these cells increases when parasites or bacteria invade the body.
- *Basophils*: The granules of these cells get stained with basic dyes, and the nucleus is more or less twisted. They form 1% of the total number of leucocytes. The cytoplasmic granules contain heparin, suggesting their anticoagulating function.
- *Neutrophils*: The granules of these cells are stained with neutral dyes. The nucleus is twisted and more than two

lobed. They form about 70% of the total number of leucocytes. They are phagocytic in function, forming the first line of defence against germs (bacteria, viruses etc.).

Agranulocytes

They usually contain clear cytoplasm and a nucleus which is not lobed. They are of two types, namely:

- Monocytes, which are large and show a kidney-shaped nucleus. They are phagocytic in function. They are about 1 to 5% of the total number of leucocytes.
- Lymphocytes, which are small with a comparatively large spherical nucleus in each. They produce antibodies and antitoxins highly specific in their reactions. They form about 25% of total leucocytes.

Blood Platelets

Platelets are minute and oval discs about 2mm in diameter. They are formed by the fragmentation of large cells called megakaryocytes of bone marrow. They are about two hundred and fifty thousand to three hundred thousand per cubic millimeter. When they come in contact with a damaged vascular surface, they release a substance called thrombokinase which initiates blood clotting. They swell, become irregular in shape and sticky and accumulate at the break of the vessel forming a platelet plug to close the opening.

Preparation of Blood Smear

- Clean one of the fingers of left hand with a cotton soaked in alcohol.
- Prick the tip of the cleaned finger with a sterile needle and squeeze the finger tip to obtain a drop of blood.
- Place this drop of blood at one end of a clean dry microscope slide.
- Touch the drop of blood with the edge of dry slide at an angle. The drop of blood spreads along the sides of the slide.
- Move the edge of the slide forward to get a thin film of blood smear.

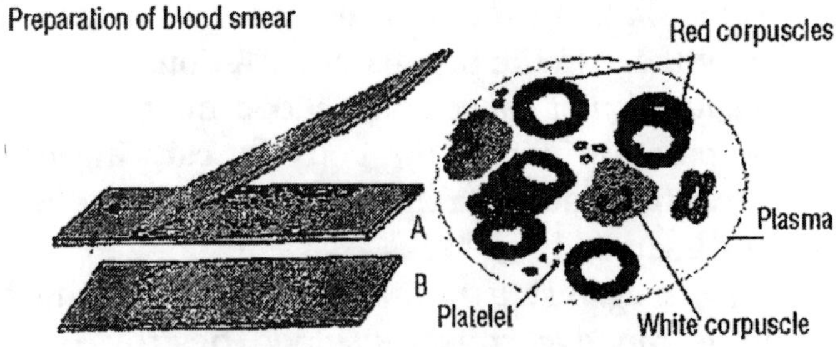

Fig. Preparation of Blood Smear and Observation of Blood Cells

- Leave the slide to dry. Pour 70% of alcohol on the smear for fixing and dry again.
- Stain the slide with Leishman's blue or Wright's stain. After a minute or two, wash the slide with water to remove the extra stain.
- Dry the slide again and observe directly under microscope.

Functions of Blood

The importance of blood is proved by its manifold functions:

- *Prevention of desiccation:* Blood supplies an adequate quantity of water to cells and prevents the tissue from dehydration. However dry the skin may be blood always keeps the interior of the body supplied with water.
- *Distribution of food:* Blood distributes back digested and absorbed food from the alimentary canal to all parts of the body.
- *Transportation of respiratory gases:* Blood carries oxygen from the lungs to the different tissues in the body and carbon dioxide from the body tissues back to the lungs.
- *Transmission of hormones:* Blood conveys hormones, the secretion of ductless glands, from the glands to the tissues and organs whose metabolism and functions they control.
- *Removal of waste products:* Blood washes out waste

materials like urea and uric acid from the cells and carry them to the organs of excretion.

- *Equalization of temperature:* Blood maintains the body temperature within the body by carrying heat from active tissues to extremities (skin, lungs) where it can be dissapated.
- *Defence against parasitic invasion:* With the help of white blood corpuscles, blood destroys germ cells and protects our body from diseases.
- *Chemical deliberation:* Blood brings about many chemical transformations, the chief being the formation of antibodies which often confines a further attack of a particular disease.
- *Clotting and healing the wounds:* Bloods stops excessive wastage of nutrients from cuts and wounds by forming a clot. The leucocytes often help in healing the wounds. It is believed that a few of them are converted into connective tissue elements of the repaired region.

Because components of blood change in response to changes in the body (diseases, infection, malfunction), blood chemistry is useful for clinical diagnosis.

Chapter 8

Photosynthesis

Energy is needed for existence as well as maintenance of life. Solar radiation is the unending source of energy for all forms of life on earth. However, living organisms can use only one form of energy for various life activities i.e. the biologically usable form of chemical energy.

It is represented by the energy-rich organic compound called ATP (adenosine tri-phosphate). Photosynthesis is the only process in which the solar energy is trapped and converted into the chemical energy ATP. This process is described as photophosphorylation. This energy is stored in food molecules as food energy. In respiration, organic food is oxidized and the food energy is released as free energy and is made available for various cellular activities. In cells, the ATP molecules act as energy mediators.Organisms which can synthesize their own organic food from simple inorganic nutrients are called autotrophs or producers. These are mostly photosynthetic (e.g. green plants) which take energy from light.

A few are chemosynthetic (e.g. certain soil bacteria) which obtain energy from chemicals. Photosynthesis is the most important and essential physiological process in green plants. This is because it is the primary source of organic food and ATP energy for all forms of life. It utilizes carbon dioxide and converts it into organic food (carbohydrates). It also releases oxygen into the atmosphere. In fact, oxygenic photosynthesis was responsible for converting the totally anaerobic atmosphere on earth into the aerobic atmosphere that exists today.

CHLOROPLASTS

Chloroplasts are green plastids. These are the special protoplasmic organelles present only in the green cells of plants. Chloroplasts act as photosynthetic apparatus. The entire process of photosynthesis is completed in each chloroplast. Hence these are the site of photosynthetic reactions. Leaves are specialized photosynthetic organs and hence, they contain the maximum number of chloroplasts in their mesophyll cells. Moreover the morphology and the anatomy of leaves are most helpful during photosynthesis for:

- Getting maximum absorption of sunlight,
- Providing steady exchange of water to green cells, and
- Allowing free exchange of carbon dioxide and oxygen.

ULTRASTRUCTURE OF CHLOROPLAST

The chloroplasts in higher plants are microscopic and mostly oval, spherical or discoid. Each chloroplast is bounded by two smooth and selectively permeable cytoplasmic membranes with an inter-membrane space. These are composed of lipo-protein subunits. The internal space of the chloroplast is filled with a colourless hydrophilic matrix called the stroma. A number of grana are suspended in the stroma. Each granum is a stack (compact bundle) of thylakoids. These are membrane-bound, flattened, disc-shaped vesicles. The thylakoid membranes are called grana lamellae. All grana are connected to one another by stroma lamellae (i.e. inter-grana lamellae or frets). The internal space of each thylakoid is called a fret channel.

The thylakoid lamellae are composed of alternating layers of lipids and aqueous proteins. There is a layer of chlorophyll and carotenoid molecules situated between the protein and lipid layers. The chlorophyll molecules are arranged in such a way that their hydrophilic heads extend into the aqueous protein layer while the lipophilic tails are embedded in the lipid layer. The pigments are organized into numerous photosynthetic units called quantasomes.

Each quantasome contains about 230 to 300 chlorophyll molecules. Quantasomes are capable of trapping light energy and converting it into chemical energy (ATP) during the photochemical reactions (light reaction) of photosynthesis. The grana also contain various co-enzymes and electron acceptors necessary for the process. Hence, grana are the site of the light reactions (phase-I) in photosynthesis.

The stroma contains various enzymes required for the dark reaction(i.e., the biochemical reactions involving the reduction of carbon dioxide to form carbohydrates). Hence, stroma is the site of the dark reactions (phase-II) of photosynthesis. The "dark reactions" do not directly require light, but they do require energy produced from light during the light reactions. The grana thylakoids and the stroma lamellae together form an intricate internal membrane system in the chloroplast. This system is derived from the inner limiting membrane during the development of chloroplast.

Pigments in Photosynthesis

The most common photosynthetc pigments present in higher plants and green algae are:

- Chlorophyll-a (blue-green) = $C_{55}H_{72}O_5N_4Mg$
- Chlorophyll-b (yellow-green) = $C_{55}H_{70}O_6N_4Mg$
- Carotenoids - Carotenes (orange-red) = C_{40} H_{56}
- Xanthophylls (yellow) = $C_{40}H_{56}O_2$

For photosynthesis, these pigments can absorb and use light belonging to the visible spectrum only. Both chlorophyll-a and chlorophyll-b show maximum light absorption in the blue-violet and in the red regions of the visible range of wavelengths of light.Carotenoids absorb light in the blue and blue-green regions. They also protect the chlorophyll from undergoing photo-oxidation when exposed to very high intensity light.

Chlorophyll-a is the essential pigment in photosynthesis, because only chlorophyll-a can utilize the absorbed light energy for the synthesis of chemical energy ATP. Other pigments act as accessory pigments. They collect the light energy and transfer it to chlorophyll-a for photosynthesis.

THYLAKOIDS IN PROKARYOTES

In prokaryotes like cyanobacteria, purple bacteria, etc., thylakoids are present but they lie naked in the cytoplasm. Chloroplasts are absent. In prokaryotes, pigments are distributed uniformly on or in the lamellae.

OVERALL EQUATION OF PHOTOSYNTHESIS

Photosynthesis is the intracellular anabolic process, characteristic of the green cells of plants in which carbohydrates are synthesized from carbon dioxide and water in the presence of light and chlorophyll. In this process, light energy is converted into chemical energy and stored in carbohydrate molecules, while oxygen is liberated.

The overall process of photosynthesis is represented by the following general chemical equation.

$$6CO_2 + 12HL2O \xrightarrow[chlorophyll-a]{light} C_6H_{12}O_6 + 6H_2O + 6O_2$$

Primary Processes of Photosynthesis

Nature of light

As stated earlier, solar radiation is the only natural source of light for all organisms. Green plants can utilize light belonging to the visible spectrum only for photosynthesis. This visible part of the electromagnetic spectrum from the sun is confined between wavelengths of 390 nm (violet) and 760 nm (red).

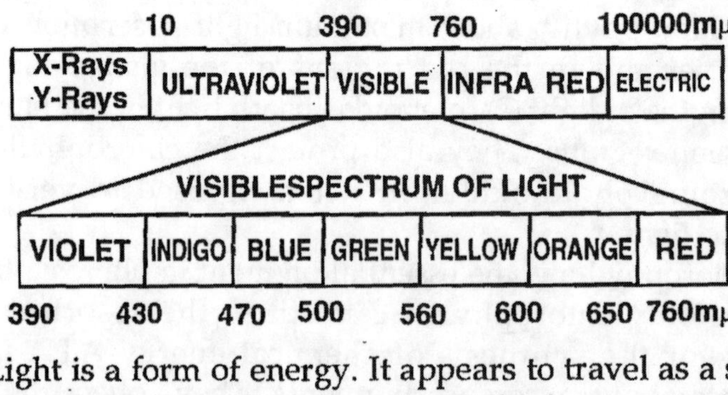

Light is a form of energy. It appears to travel as a stream of discrete particles called photons. Each photon contains one quantum (unit) of light energy. The energy quantity of each type of light depends on its wavelength. Light with a shorter wavelength has greater energy; light with a longer wavelength has lesser energy.

Rate of photosynthesis is measured in terms of the amount of carbon dioxide used (reduced) or oxygen released. It has been estimated that 8 quanta (photons) of light are required to reduce each molecule of carbon dioxide (or to release each molecule of oxygen) during photosynthesis. This is called the quantum requirement in photosynthesis.

Light trapping systems: During the photochemical phase, light is trapped by the photosynthetic pigments present in the quantasomes of the grana thylakoids. These pigments are organized into two pigment systems called pigment system I (PS I) and pigment system II (PS II).

- *PS I is composed of the following pigment molecules:*
  - Chl-a 700 (P700) - one molecule - Reaction centre
  - Chl-a 683 - 200 molecules - Antenna chlorophyll
  - Carotenoids - 50 molecules - Accessory pigments
- *PS II is composed of the following pigment molecules:*
  - Chl- a 680 - one molecule - Reaction centre
  - Chl- a 670 - 200 molecules - Antenna chlorophyll
  - Chl- b - up to 200 molecules} Accessory pigments
  - Carotenoids - up to 50 molecules

In both the systems, all pigment molecules help in trapping light energy. However, all other molecules transfer their energy to the reaction centre to be used for the photochemical reactions in photosynthesis.

Photo-excitation of Chlorophyll-a

When a molecule of chlorophyll-a (acting as the reaction centre) receives light energy in the form of photons (quanta), it becomes energy-rich and activated. This is called the excited state of chlorophyll. In this excited state, the chlorophyll-a expels one electron. With the loss of an electron, the chlorophyll-a develops a positive charge. This is called ionized chlorophyll-a. The expelled electron contains an extra amount of energy received from light. In other words, the energy-rich expelled electron represents light energy. This electron energy is used for the formation of ATP during photosynthesis.

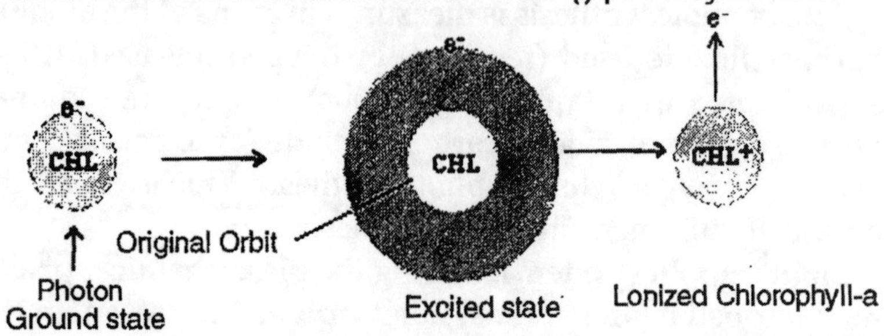

Fig. Phto-excitation of Chloropyll-a and the Loss of an Electron.

Photolysis of Water

In the photosynthesis of green plants, water is used as a source of hydrogen required for the reduction of carbon dioxide to form carbohvdrates.

$$CO_2 + 2H_2O \xrightarrow[\text{Chlorophyll}]{\text{light}} \underset{\text{carbohydrate}}{(CH_2O)} + H_2O + O_2$$

In this process, water is oxidized in presence of light and chlorophyll. Hydrogen is removed from water, and oxygen is released as follows:

$$4H_2O \xrightarrow[\text{Chlorophyll}]{\text{light}} 2H_2O + 4H^+ + 4e^- + O_2$$

$$\text{OR } 2H_2O \longrightarrow 4H^+ + 4e^- + O_2$$

This is described as photolysis of water, or photochemical oxidation of water. This was first indicated by Van Niel. It was demonstrated with the help of an experiment by R. Hill. Hill suggested that during the photolysis of water, hydrogen combines with some hydrogen acceptor in the plant, while oxygen is released.

$$2H_2O + 2A \xrightarrow[\text{Chlorophyll}]{\text{light}} 2AH_2 + O_2$$

where A = unknown H-acceptor in plants).

Later on, Arnon discovered this hydrogen acceptor to be a coenzyme called NADP (nicotinamide adenine dinucleotide phosphate).

Substance A = NADP

$$2H_2O + 2NADP \xrightarrow[\text{Chlorophyll}]{\text{light}} 2NADPH_2 + O_2$$

reduced co-enzyme

Similarly, Ruben, Kamen, et al. confirmed that the oxygen evolved during photosynthesis comes from the splitting of water only. They used heavy isotope of oxygen (O_{18}) to demonstrate the fact as follows:

$$\text{(I)}\ 6CO^{18}{}_2 + 12H_2O \xrightarrow[\text{Chlorophyll}]{\text{Light}} C_6H_{12}O^{18}{}_6 + 6H_2O^{18} + 6O_2$$

$$\text{(II)}\ 6CO_2 + 12H_2O^{18} \xrightarrow[\text{Chlorophyll}]{\text{light}} C_6H_{12}O_6 + 6H_2O + 6O^{18}{}_2$$

HILL REACTION

R. Hillsuspended isolated chloroplasts in water in a beaker. He added some known hydrogen accepting compound (e.g. benzoquinone or ferric salt) to the water and exposed the beaker to the light. Immediately the oxidation of water took place and bubbles of oxygen were evolved. When the hydrogen acceptor in the beaker was examined, it was found to be reduced.

$$2H_2O + 2A \xrightarrow[\text{Chlorophyll}]{\text{light}} 2AH_2 + O_2$$

A = hydrogen acceptor).

This is photolysis or the photochemical oxidation of water and is commonly called the Hill reaction.

Photophosphorylation

The formation of ATP molecules (chemical energy) from ADP and H_3PO_4 in presence of light and chlorophyll-a during the photochemical phase of photosynthesis is called photophosphorylation or photosynthetic phosphorylation. More commonly, it is described as the conversion of light energy into chemical energy (ATP).

ADP = adenosine diphosphate)

During the light reactions (the primary process or the

photochemical phase), ATP formation takes place through two types of phosphorylation reactions.

$$ADP + H_2PO_4 \xrightarrow[\text{Chlorophyll-a}]{\text{Light}} ATP$$

These are

- Non-cyclic photophosphorylation and
- Cyclic photophosphorylation

The non-cyclic process involves both PS I and PS II. The cyclic process involves only PS I.

NON-CYCLIC PHOTOPHOSPHORYLATION

In the light reaction of photosynthesis, the formation of ATP from ADP and H_3PO_4 in presence of light and chlorophyll-a during the non-cyclic transfer of electrons is called non-cyclic photophosphorylation.

Important Features

- It is the major pathway of the light reaction of photosynthesis.
- It is a photochemical reaction.
- It takes place in the grana of chloroplasts.
- It requires participation of both PS I and PS II.
- The transfer of electrons through the ETS is unidirectional or non-cyclic.
- The non-cyclic process involves:
  - Photophosphorylation (ATP formation)
  - Photolysis of water
  - Formation of assimilatory power (ATP and $NADPH_2$)
  - Liberation of oxygen

The non-cyclic photophosphorylation takes place as follows.

Photoexcitation of PS I

The chlorophyll-a (P 700) of PS I is activated on receiving photons of light and so it expels electrons. As a result of this loss, it becomes ionized chl-a$^+$.

The electrons from PS I are first accepted by FRS (ferredoxin reducing substance) and are transferred to co-enzyme NADP via Fd (ferredoxin). NADP retains the electrons and is thus reduced.

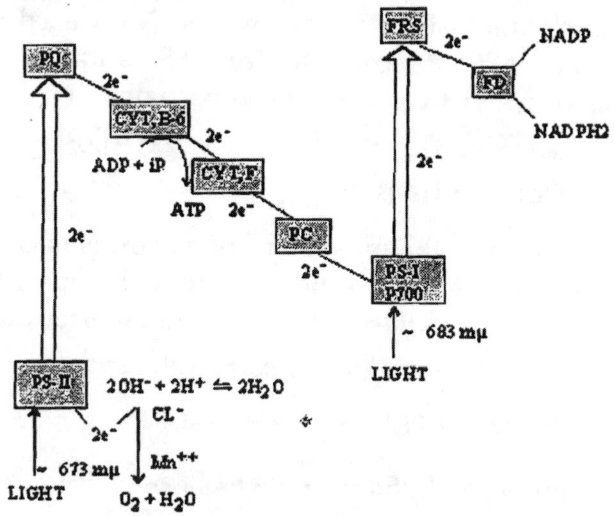

Fig. Diagrammatic Representation of the Non-cyclic Process (Z scheme)

Photoexcitation of PS II

Similarly, on receiving photons of light, the chl-a (680) of PS II also gets activated and expels electrons. As a result, it becomes ionized chl-a^+. The high energy containing electrons from PS II are first accepted by an electron acceptor PQ (plastoquinone) and are then transferred along stet Electron Transport System consisting of plastoquinone, cytochrome b6, Cytochrome f and plastocyanin. Finally, the electrons are accepted by the ionized chl-a^+ of PS I. In this way, ionized PS I comes back to the ground state and can again participate in the process.

Photolysis of Water

$$2H_2O \xrightarrow[\text{chlorophyll}]{\text{light}} 4e^- + 4H^+ + O_2 \text{ (Photolysis of water)}$$

The ionized chlorophyll-a^+ of PS II is brought back to the ground state with the help of the electrons made available through the photolysis of water.

FORMATION OF NADPH2 AND LIBERATION OF O_2

During photolysis of water, oxygen is released as a by-product. The 2 NADP receive 4e- from PS I and $4H^+$ from the photolvsis of water to form reduced coenzvme $NADPH_2$

$$2\,NADP + 4e^- + 4\,H^+ \longrightarrow 2NADPH_2$$

Non-Cyclic Electron Transfer

In this process, the movement of electrons is considered to be non-cyclic, or unidirectional. This is because, the final electron acceptor is different from the initial electron donor.

Initial Electron Donor Final Electron Acceptor

(I) Chl-a of PSI $\xrightarrow{e^-}$ NADP

(Ii) Chl-a of PSII $\xrightarrow{e^-}$ Chl.a+ of PSI

(Iii) Water (Photolysis) $\xrightarrow{e^-}$ Chl.a + of PS II

In this way, the light-induced electron flow is maintained continuously in the non-cyclic manner from water to NADP as fc¨

$$\text{Water} \xrightarrow{e^-} \text{PSII} \xrightarrow{e^-} \text{PS I} \xrightarrow{e^-} \text{NADP}$$

Because of the zig-zag path of the electrons, the non cyclic process is also called 'Z' scheme electron transport.

ATP Formation

During the non-cyclic transfer of energy-rich electrons through the Electron Transport System from PS II to PS I, energy from the electrons is released. This is used for the formation of ATP from ADP and H_3PO_4. This is called non-cyclic photophosphorylation.

End Products of Non-cyclic Process and the Assimilatory Power

The end products are ATP, $NADPH_2$ and O_2. Of these, O_2 is liberated from the green plants. ATP and $NADPH_2$ are used in the following dark reactions (phase II) for carbon assimilation(i.e., reduction of CO_2 to form carbohydrates). ATP and $NADPH_2$ of light reaction are called 'assimilatory power.

CYCLIC PHOTOPHOSPHORYLATION

In the light reaction of photosynthesis, the formation of ATP from ADP and H_3PO_4 in the presence of light and chlorophyll-a during the cyclic transfer of electrons is called cyclic photophosphorylation.

Important Features

- It is a part of the light reaction of photosynthesis.
- It is a photochemical reaction and depends on light.
- It takes place in the grana of chloroplasts.
- It requires participation of PS I only.
- The movement of electrons is cyclic.
- The cyclic process involves only ATP formation.

The various steps involved are as follows:

Photoexcitation of PS I

The reaction centre of PS I is activated on receiving light energy in the form of photons. In this excited state, the chl-a of PS I expels an electron and becomes ionized chl-a.

Cyclic Electron Transfer

The energy rich electrons from the PS I are first accepted by FRS and transferred to Fd. From Fd, the electrons are transferred to Cytochrome. b6 (and not to NADP), Cytochrome. f and then to PC. From PC the electrons are finally accepted by the same ionized chl-a^+ of PS I. On receiving the lost electrons, the chlorophyll-a^+ comes back to the ground state and can again participate in the process. In this process, the initial donor and the final acceptor of electrons is the same chlorophyll-a of PS I. Hence, the movement of electrons is cyclic.

Synthesis of ATP

During the cycle transfer of the energy-rich electrons, there is loss of electron energy at each transfer. It is used at one such transfer for the formation of ATP from ADP and H_3PO_4. This is called cyclic photophosphorylation. In the cyclic process, neither PS II or the photolysis of water are involved. Hence, $NADPH_2$ is not formed and oxygen is not evolved.

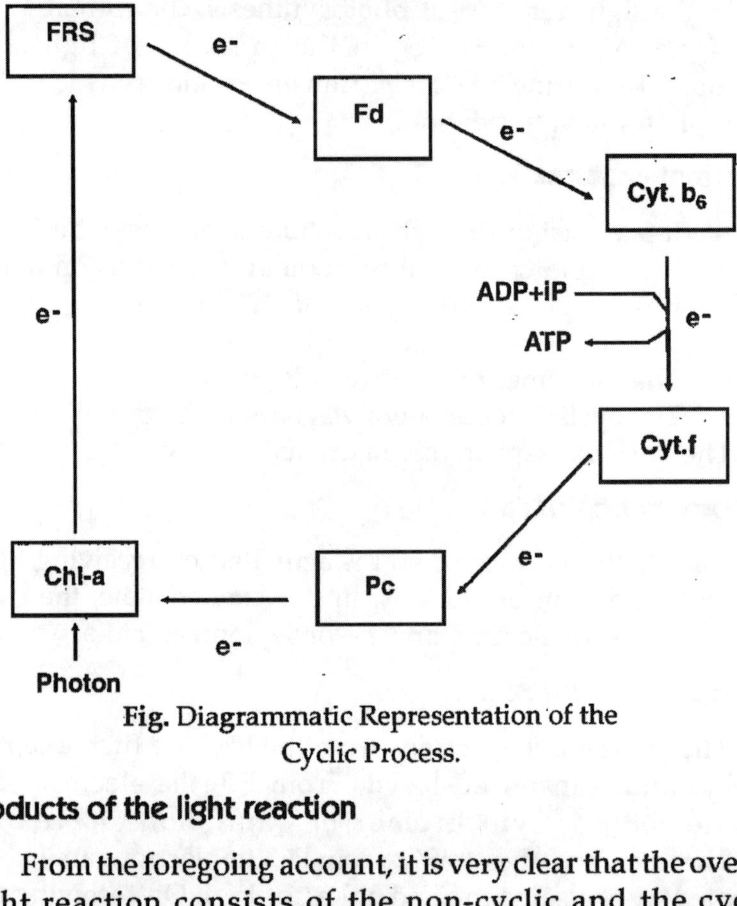

Fig. Diagrammatic Representation of the Cyclic Process.

Products of the light reaction

From the foregoing account, it is very clear that the overall light reaction consists of the non-cyclic and the cyclic photophosphorylation reactions.

The products of the light reaction (cyclic + non-cyclic) are ATP, $NADPH_2$ and O_2. As stated earlier, the oxygen is liberated from the green plants. ATP and $NADPH_2$ are used in the following dark reactions (phase-II) for the reduction of cabon dioxide to form carbohydrates (carbon assimilation). Hence, these products of the light reactions (ATP and $NADPH_2$) are called assimilatory power (Arnon).

SECONDARY PROCESSES OF PHOTOSYNTHESIS (BIOCHEMICAL PHASE, DARK REACTIONS)

The second part or phase-II of photosynthesis is known as the dark reactions. This is because, the reactions in this phase do not require light. These reactions are thermo-chemical and depend on temperature. The dark reaction involves thermo-chemical reduction of carbon dioxide to form carbohydrates. This was first established by Blackmann, hence it is also called Blackmann reaction. Presence of light is not necessary for the reduction of carbon dioxide. However, the dark reaction utilizes the products of light reaction (i.e., the assimilatory power (ATP and $NADPH_2$) to reduce carbon dioxide into carbohydrates.)

All the reactions in phase-II are completed in the stroma of the chloroplast. These are biochemical reactions in which every step is controlled by a specific enzyme. The reactions begin with the fixation of carbon dioxide, involve the utilization of the assimilatory power for the reduction of carbon dioxide and end with the formation of carbohydrate, which is the end product of photosynthesis. The details of the steps involved in the dark reaction (i.e., the complete path of carbon) were worked out by Professor M. Calvin and hence, the dark reaction came to be called as Calvin cycle.

CALVIN CYCLE (C_3 PATHWAY)

Important features

- This is the major pathway for the fixation of carbon dioxide in green plants.
- It represents phase-II, i.e., the dark reaction of photosynthesis.
- It takes place in the stroma of the chloroplasts.
- The reactions are enzyme-controlled and temperature dependent.
- It was discovered by Professor M. Calvin and therefore called the Calvin cycle.
- After the fixation of carbon dioxide, the first stable compound formed is 3-carbon phosphoglyceric acid (PGA). Hence it is also called the C_3 pathway.

The important events of the Calvin cycle can be studied under the following heads.

The first CO_2 acceptor and fixation of CO_2

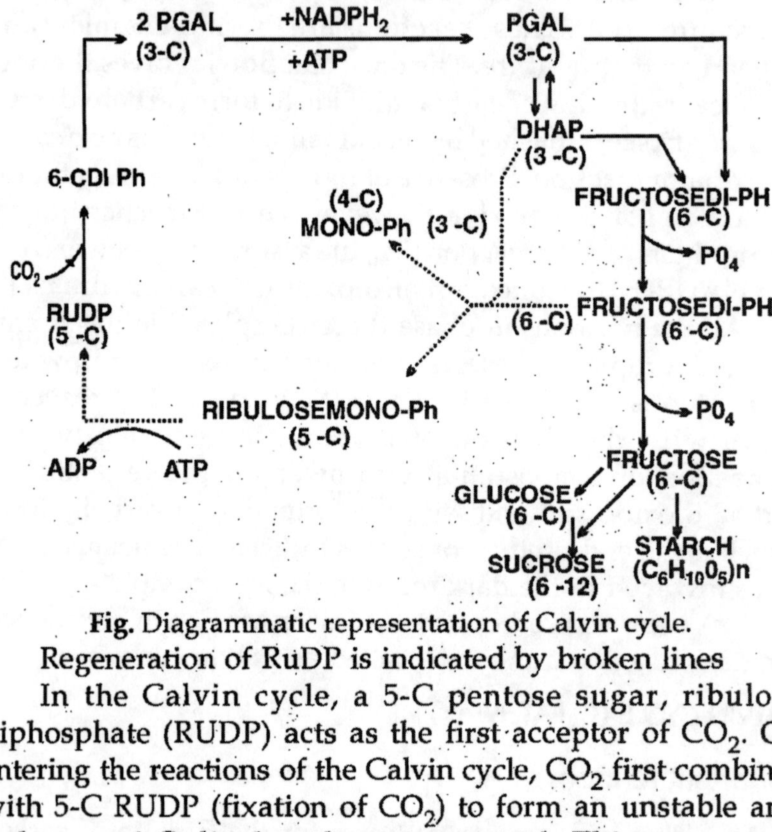

Fig. Diagrammatic representation of Calvin cycle. Regeneration of RuDP is indicated by broken lines

In the Calvin cycle, a 5-C pentose sugar, ribulose diphosphate (RUDP) acts as the first acceptor of CO_2. On entering the reactions of the Calvin cycle, CO_2 first combines with 5-C RUDP (fixation of CO_2) to form an unstable and unknown 6-C disphosphate compound. This compound immediately breaks into 2 molecules of 3-C phosphoglyceric acid (PGA). The enzyme carboxy dismutase catalyzes the reaction. Thus, 3-C PGA is the first stable compound formed in Calvin cycle. Hence, it is also called C-3 pathway.

Utilization of assimilatory power (or reduction of PGA)

The 3-C PGA then undergoes reduction with the help of the assimilatory power to form 3-C phosphoglyceraldehyde (PGAL). $NADPH_2$ provides the hydrogen and ATP supplies energy for the reduction. Enzyme triosephosphate dehydrogenase catalyzes the reaction. Some molecules of

PGAL are converted into another triose phosphate called dihydroxyacetone phosphate (DHAP) in presence of enzyme triose phosphate isomerase.

Formation of sugars (end products of photosynthesis)

The 3-C triose phosphates(i.e., PGAL (3-C) and DHAP (3-C)) form 6-C hexose sugar fructose.1, 6 diphosphate in the presence of enzyme aldolase. Fructose diphosphate is then dephosphorylated first to fructose mono-phosphate and then to fructose (6-C) in the presence of enzyme phosphatase. Some fructose monophosphate molecules may be isomerized into glucose monophosphate by enzyme isomerase and then into glucose (6-C). The hexose sugars may be further converted to sucrose ($C_{12}H_{22}O_{11}$) or to starch ($(C_6H_{10}O_5)n$) and stored in storage cells.

Regeneration of the CO_2 acceptor, RUDP

The 5-C RUDP is constantly required for the fixation of CO_2 in the Calvin cycle. It is regenerated through another chain of reactions. Some molecules of triose phosphates and fructose mono phosphates are used from the Calvin cycle for the formation of RUDP.

DIVERSITY IN PHOTOSYNTHETIC PATHWAY

Various experiments and investigations regarding the fixation of CO_2 and the path of carbon during the dark reaction have indicated that there are a number of different pathways for CO_2 fixation in green plants. These are

- C_3 Pathway (Calvin cycle)
- C_4 Pathway (H-S pathway or Hatch-Slack cycle)
- CAM (Crassulacean acid metabolism)

The C_3 pathway

The C_3 pathway or Calvin cycle is the main pathway and is present in most green plants. Plants which follow the Calvin cycle are called C_3 plants.

C_4 Pathway (or Hatch-Slack pathway)

In some plants, the first stable product, after the fixation of CO_2, is a 4-C dicarboxylic acid called oxaloacetic acid (OAA).

Such plants are called C_4 plants and the path of carbon (dark reaction) is the C_4 pathway. It was first noticed by Kortschak in the photosynthesis of sugarcane leaves. However, details of the C_4 pathway (i.e. the first CO_2 acceptor, the first stable product, the complete carbon pathway, etc.) were worked out by Hatch and Slack. Therefore, it is known as Hatch-Slack Pathway. C_4 pathway is observed in many plants of family Gramineae (e.g. sugarcane, maize, some other monocot and some dicot plants).

Anatomical peculiarities of C_4 plants

Most of the C_4 plants have a characteristic leaf anatomy and dimorphic chloroplasts. For example,

- The leaf mesophyll consists of more or less compactly arranged cells.
- It is not differentiated into palisade and spongy mesophyll as it is in C3 plants.
- The vascular bundles (veins) in the leaf are surrounded by a distinct bundle sheath of radial enlarged parenchyma cells.
- The chloroplasts in leaf cells are dimorphic(i.e., of two types):
  - Chloroplasts in the mesophyll cells are smaller and possess grana.
  - Chloroplasts in the bundle sheath cells are larger and without grana.

This type of leaf anatomy in C_4 plants is described as Kranz anatomy.

Important steps in Hatch and Slack Pathway

The various steps in C_4 pathway are completed in two parts and in two different regions in the leaves. (a) First part reactions are completed in the stroma of the chloroplasts in mesophyll cells, and (b) Second part reactions are completed in the stroma of chloroplasts in bundle sheath cells.

Part - I (in mesophyll cells)

- First CO_2 Fixation: In the pathway, the first CO_2 acceptor is the 3-C phosphoenol pyruvate acid (PEP).

CO_2 first combines with 3-C PEP to form 4-C OAA (oxaloacetic acid). As OAA is a dicarboxylic acid, this is also known as the dicarboxylic acid pathway.

- 4-C OAA may be converted into 4-C malic acid or 4-C aspartic acid and transported to bundle sheath cells.

Part - II (in bundle sheath cells)

- In the chloroplasts of the bundle sheath cells, 4-C malic acid undergoes decarboxylation to form CO_2 and 3-C pyruvic acid.

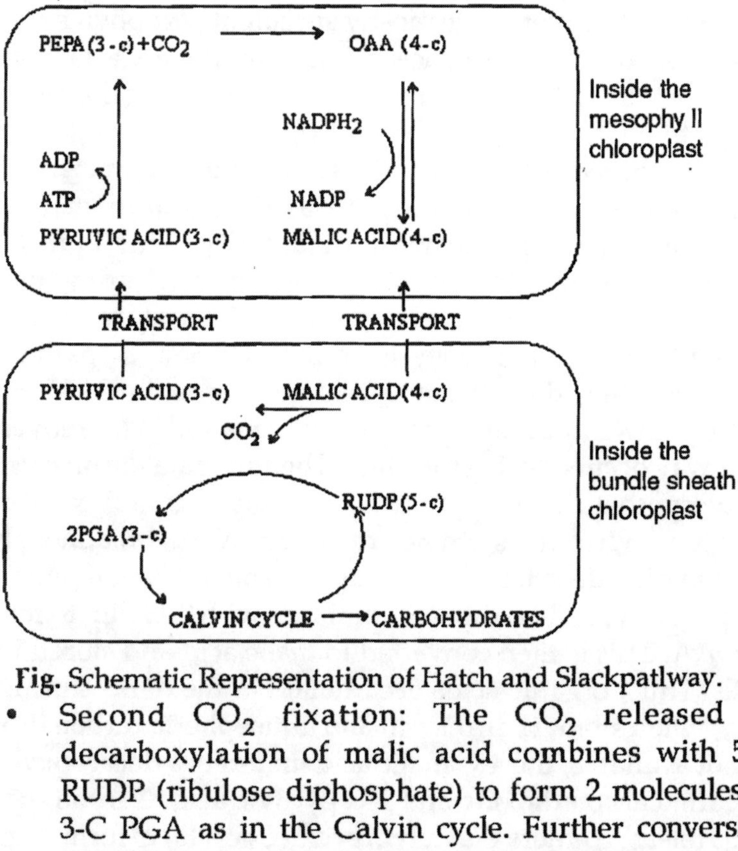

Fig. Schematic Representation of Hatch and Slackpatlway.

- Second CO_2 fixation: The CO_2 released in decarboxylation of malic acid combines with 5-C RUDP (ribulose diphosphate) to form 2 molecules of 3-C PGA as in the Calvin cycle. Further conversion of PGA to sugars is the same as in the Calvin cycle.
- The pyruvic acid produced in decarboxylation of malic acid is transported back to the mesophyll cells. Here it is converted into PEPC and again made available for the C_4 pathway.

Thus, in the C_4 plants, the initial few steps are different (typical of C_4 pathway). However, later on the reactions are similar to the Calvin cycle (i.e., C_3 pathway). Hence, C_4 plants have both C_4 and C_3 pathways. Moreover, in C_4 plants, CO_2 fixation (i.e. carboxylation) occurs twice: first in the mesophyll cells (PEP + CO_2) and then again in the bundle sheath cells (RUDP + CO_2). For this reason, the C_4 pathway is also called a dicarboxylation pathway.

CAM plants (Crassulacean acid metabolism)

The CAM plants are mostly succulent xerophytes such as plants of families Crassulaceae, Euphorbiaceae, Cactaceae, etc. The stomata in these plants remain closed during the day. This helps to check transpiration and open them during the cool night for gaseous exchange. These plants are characterized by the presence of Crassulacean acid metabolism (CAM) involving night photosynthetic activity. This is a desert adaptation where stomata must remain closed during the day.

The CAM plants have the dark reaction very much similar to the C_4 plants (i.e., they have both C_4 and C_3 pathways.) However, they differ from C_4 plants because, in CAM plants, both pathways occur in mesophyll cells only. Moreover, C_4 pathway occurs during the night when stomata are open (night photosynthetic activity) and C_3 pathway is completed during day time when the stomata are closed. At night, atmospheric carbon dioxide is taken in through the open stomata. As in the C_4 pathway, carbon dioxide combines with 3-C PEP to form 4-C OAA. This is then converted to malic acid and stored in the cells. Thus, organic acids accumulate in the dark. During the day, when stomata are closed and atmospheric carbon dioxide is not available, the 4-C malic acid undergoes decarboxylation to form carbon dioxide and 3-C pyruvic acid. Carbon dioxide thus released enters the Calvin cycle (C_3 cycle) to form sugars.

SIGNIFICANCE OF PHOTOSYNTHESIS

- It is the primary source of organic food and food energy (ATP) for all forms of life, either directly or indirectly.

- Excess sugars produced in photosynthesis are either stored in the form of carbohydrates or used in the biosynthesis of other organic compounds.
- In any ecosystem, green plants represent the most essential biotic components as they are the primary producers.
- Photosynthesis helps to purify air and also maintain balance of oxygen and carbon dioxide in the ecosystem.
- Oxygenic photosynthesis was responsible for converting the totally anaerobic condition on earth into aerobic atmosphere present now.
- The fossil fuels (e.g. natural gas, coal, petroleum (oil), etc.) are all energy-rich materials of an organic origin. The energy stored in all these fuels is basically solar energy which was trapped and stored during photosynthesis in the geological past.

Chapter 9

Viruses

A five-kingdom classification system fails to classify viruses because they are not considered to be living. They neither reproduce independantly nor utilize energy. Viruses are now defined as ultramicroscopic disease-producing entities living in a host as obligatory intracellular parasites.

Viruses were not known to biologists for a long time due to thier ultramicroscopic structure although its presence was apparent by infectious diseases which were not due to bacteria. They attracted the attention of investigators in the 19th century when mosaic disease (TMV) caused severe damage to commercially important tobacco crop.

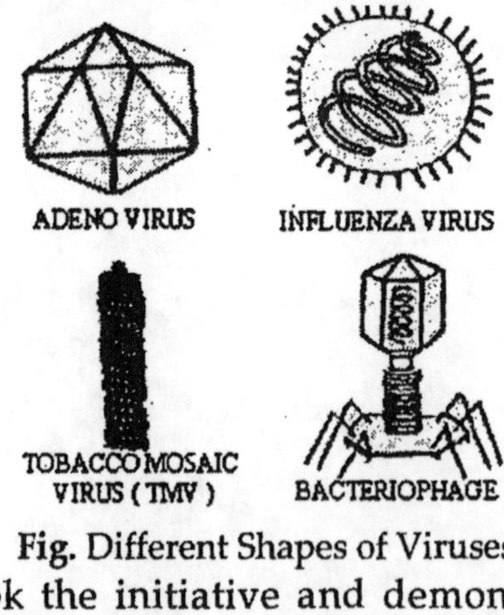

Fig. Different Shapes of Viruses

Mayer took the initiative and demonstrated that the disease could be transmitted just by applying the sap of infected leaves to the leaf of healthy plant. Mayer thought that the causative agent was a bacterium. A crucial experiment was performed by Russian biologist Iwanowsky who demonstrated that the sap of infected leaves, even after passing through the finest porcelain filter (which prevents bacteria to pass through), remained infective. He, therefore, ruled out the presence of bacterium but could not think of any other causative agent.

Dutch microbiologist Beijerinck onfirmed Iwanowsky's findings and suggested that the causative agents were not any submicroscopic particles but the fluid itself which he called "contagium vivum fluidum" which means contagious living fluid. This was later on called viron (meaning poison) and the disease causing agent called virus. Twort and d' Herelle independently discovered viruses attacking bacteria for the first time.

After the discovery of the ultramicroscope, the controversy was finally laid down to rest when W. M. Stanley, the American biochemist, isolated the virus in crystalline form and showed that even in this state it retains infectivity. These findings gave inspiration to other scientists all over the world to study more aspects of viruses. This marked the beginning of a new branch of science called virology.

GENERAL CHARACTERS

A large number of viruses are now known. They exhibit diversity of form and infect a number of organisms. Despite diversity of form and structure, they show the following important characteristics common to all viruses:

- They are ultramicroscopic disease-producing entities.
- They have no cellular organization and also no metabolic machinery of their own.
- They are simple in structure, basically composed of nucleic acid wrapped up in a protein coat.
- Nucleic acid is only of one type, either DNA or RNA, but never both.
- They are obligatory intracellular parasites as they are completely inactive outside the host.
- They multiply within the host by commandeering the metabolic machinery of the host cell.
- They are specific in action, i.e. they always infect particular organ or organism.
- They are incapable of growth and division.

- They can be crystallized and even in crystalline form, they retain their infectivity.
- They are unaffected by antimicrobial antibiotics.
- They may undergo mutations.

SIZE AND SHAPE

Viruses are minute entities, even smaller than the smallest bacterium. They can be seen only under electron microscope as small particles called virons. Being minute, they are measured in millimicrons (1m m = 1/1000 m). Generally they vary from 10 mm to 300 mmð in size.

Viruses occur in three main shapes, viz.

- Polyhedral or spherical, e.g. adeno virus, herpes virus etc.
- Helical or rod-like, e.g. tobacco mosaic virus, (TMV), influenza virus, etc.
- Complex or irregular, e.g. bacteriophage, vaccinia, etc.

STRUCTURE

The virus is composed of two major parts:

- Capsid
- Nucleic acid.

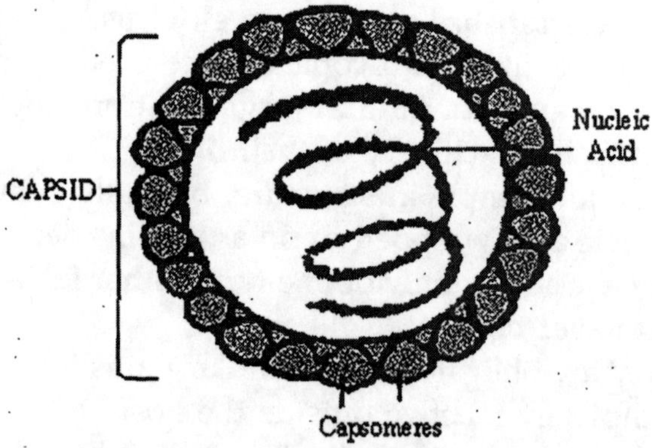

Fig. A Generalized Structure of a Virus

Capsid

The capsid is the outer protective coat mostly made up of specific protein. It protects nucleic acid from inactivation by enzyme nuclease in the environment. It is often composed of many identical subunits called capsomeres.

The shape and arrangement of capsomeres determine the shape of the virus. Some highly specialized viruses, for example. influenza virus, mumps virus etc. show an outer covering called an envelope which contains cell membrane obtained during exit of the virus from its host cell.

Nucleic Acid

The nucleic acid is in the central core. Unlike living organisms it contains a single molecule either of DNA or RNA, but never both. Nucleic acid is the only active part of a virus, hence viruses are sometimes called "wandering genes". The infectivity of virus is due to nucleic acid while host specificity is determined by the protein coat.

The capsid in close contact with nucleic acid, is known as nucleocapsid. Viruses may be enveloped or non-enveloped (naked).

Chemical Composition

A virus is a nucleoprotein, i.e. mainly consisting of nucleic acid and proteins. Nucleic acid is— either DNA or RNA, but never both. When only RNA is present, genetic information is solely carried by RNA, which is the unique property of virus. The envelope, if present, contains lipoproteins. The lipid is mostly derived from the host plasma membrane while the protein is virus coded. Viruses normally do not possess any biosynthetic enzymes.

Classification of Viruses

According to the type of hosts they infect, viruses may be classified as one of the following three types:

Plant viruses

These are pathogenic viruses which infect plants. They are usually rod-shaped, containing nucleic acid in the form of RNA, e.g. tobacco mosaic virus (TMV), beet yellow virus (BYV) etc. The capsomeres of TMV virus are elliptical and arranged helically around the central nucleic acid core.

Common plant viral diseases are:

- Mosaic disease of tobacco, papaya, apple etc.,
- Black ring spot of cabbage,
- Leaf-roll of potato,
- Spotted wilt of tomato.

Animal viruses

These are pathogenic viruses infecting animals. They are generally polyhedral or spherical in shape. The capsid in some is surrounded by envelope, and the nucleic acid is either DNA or RNA. According to the type of tissue which they infect, they are as follows:

- Dermatotropic: Viruses infecting the skin, e.g. measles, chicken pox.
- Viscerotropic: Viruses infecting viscera, e.g. yellow fever, jaundice, mumps.
- Neurotropic: Viruses infecting nervous system, i.e. polio, meningitis.

Bacterial Virus

These are pathogenic viruses infecting bacteria and are called bacteriophages or simply phages. their nucleic acid is DNA, e.g. T2, T4, T6 bacteriophages.

Significance

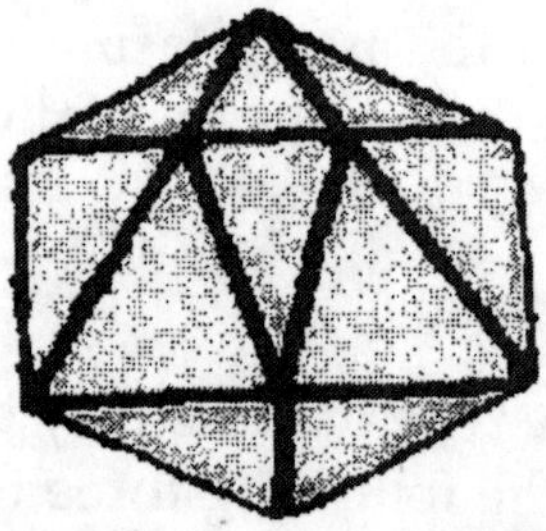

Fig. Adenovirus (animal virus)

- Viruses are a kind of biological puzzle to biologists as they are at the threshold of living and non-living, showing the characteristics of both.
- Viruses are used by humans in eradicating harmful pests like insects and in controlling the population of organisms such as rabbits by inducing viral

infection. Thus they are used as a form of biological control.

- Viruses have gained a prominent position in world because of their value as biological research tools. Due to simplicity of structure and rapid multiplication, they are widely used in research, in the fields of molecular biology, medicine and genetic engineering. Their role in fundamental research to unlock the intricate phenomena of life, can never be over- emphasized.

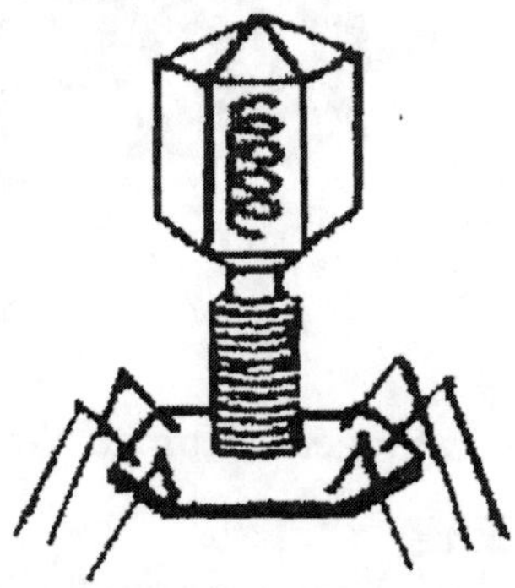

Fig. Bacteriophage (bacterial virus)

- Viruses have also concerned agriculturists. Apart from causing diseases in crops, bacteriophages attack the nitrogen fixing bacteria of the soil and are responsible for reducing fertility of the soil.
- In industry, however, viruses are used in preparation of sera and vaccines to be used against diseases like rabies, polio, etc. The multiplication of viruses in bacterial cell is also utilized in the production of antibodies.

Pathogenic Nature of Virus

Viruses are causative agents of various dreadful diseases in plants, domesticated animals and man. A number of plants like tomato, tobacco, potato, sugarcane, etc. are affected and destroyed every year by viruses. Many domesticated animals are also destroyed. The common mammalian viral diseases are foot-and-mouth diseases of cattle, encephalomyeletis of horse, distemper of dog, rabies, etc. Common human diseases caused by viruses are mumps, measles, chicken pox, small pox, herpes, influenza, common cold, jaundice, polio, etc.

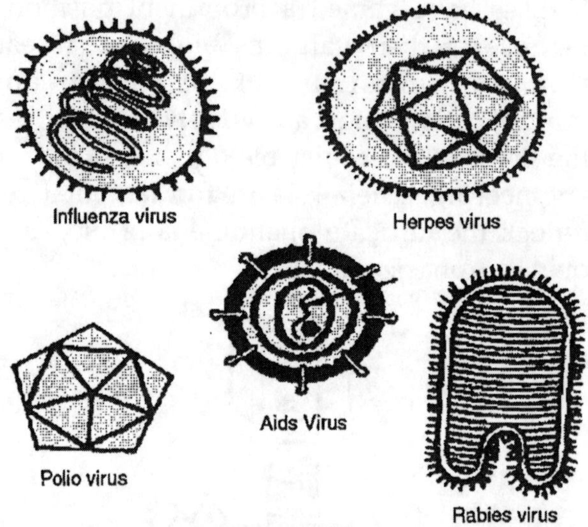

Fig. Pathogenic Human Viruses

VIRUSES AND CANCER

Cancer is an uncontrollable and unorganized growth of cells causing malignant tumors, the cells of which have the capacity to spread indiscriminately anywhere in the body.

Cancers grow by progressive infiltration, destruction and penetration of the surrounding tissues. It is curable in the initial stage but in the last stage called metastasia, the tumors break apart and the cells spread to other organs, the functions of which get disrupted, hence causing death of an individual.

In the past it was thought that cancer was not caused by viruses, but in recent years, there have been increasing evidences to prove that the cancer is caused by the DNA virus called simian virus (SV-40) and a group of RNA viruses called retroviruses. The cancer causing-viruses are also called oncogenic viruses. Cancer can be cured in its initial stages by radiation, chemotherapy and surgery. Early diagnosis greatly decrease the hazards of cancer. It is believed that viruses are involved in leukemia, sarcomas and some kinds of breast cancer.

VIRUS AND AIDS

AIDS (Acquired Immune Deficiency Syndrome) is a

recently discovered sexually transmitted viral disease. It is caused by Human Immuno Deficiency Virus (HIV), the target cell of which is T4 lymphocyte of the host, which usually forms the main line of defence. Due to failure of the human immune system, sufficient antibodies are not formed and the human body becomes susceptible to various infections.

The disease is transmitted during blood transfusion, through sexual contact, from infected mother to the child during pregnancy or through breast feeding. There is currently no cure for AIDS. The treatment of AIDS, therefore, depends upon the efforts to check the secondary opportunistic infections attacking AIDS patients, as well as slowing or halting replication of the HIV virus.

KINGDOM: MONERA

These are the oldest, simplest and most numerous microorganisms. They are distinguished by the following characters:

- They are prokaryotes.
- They are mostly unicellular, but may be in the form of colonies or filaments of independent cells.
- Their mode of nutrition is mainly absorptive, but some are photosynthetic or chemosynthetic.
- They are usually nonmotile, but some may have flagella and gliding movements. Cilia are absent.
- Their reproduction is primarily asexual, by fission.

Monera includes heterogenous microorganisms including archaebacteria, eubacteria, actinomycetes and cyanobacteria.

ARCHAEBACTERIA: (ARCHAEOS: OLD)

Fig. Archaebacteria

These are ancient bacteria which probably evolved 3 billion years ago. And are now known as "living fossils". They are biologically different from the present day bacteria in two respects, i.e. cell wall does not contain muramic acid, but it is composed of proteins and polysaccharides and the cell membrane consists of branched chain lipids. This enables them to tolerate the extremes of heat and pH. They are divided into two sub-groups.

Methanogens

These are strictly anaerobic bacteria which produce methane (CH_4) from CO_2 and formic acid, hence the name. They are present in salty, marshy places, in the stomach of cattle and in organic matter or sewage. Methane gas produced in biogas plants is due to these bacteria. They are also called halophiles as they have an affinity for salt. They can cause spoilage of salted fish. Examples: Halobacterium, Halococcus.

Sulphur- Dependent Bacteria

These are aerobic bacteria which convert sulphur either into sulphuric acid (H_2SO_4) or into hydrogen sulphide (H_2S). Hence, they are present in hot sulphur springs. They can tolerate highly acidic pH (pH=2) and high temperature. Hence, they are also called thermoacidophiles. Examples: Thermoplasma, Sulfolobus

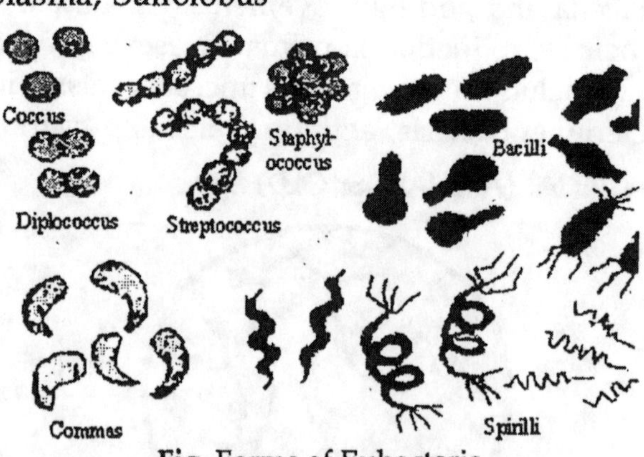

Fig. Forms of Eubacteria

EUBACTERIA

These are "true" bacteria ubiquitous in nature, i.e. they

are found practically in all the environments, at all the attitudes and depths, in extremely low and high temperature, in fresh as well as in marine water and in bodies of plants and animals both living and dead. In fact, it is difficult to name any place where bacteria are not found. They show the following general characters.

Shape and Size

They are unicellular microorganisms of various shapes and accordingly, they are called cocci (spherical), bacilli (rod-shaped), spirilla (spiral) and vibrios (broken spirals or comma shaped). Their sizes vary from 0.1m to 20 m in breadth and 0.2 m to 80 m in length.

Locomotion

They are generally non-motile, but motile bacteria may have flagella at one end, at both ends, or all around the cell.

Cell Structure

Their cell wall contains peptidoglycan and muramic acid. Their cytoplasm is without streaming movements and without endoplasmic reticulum. However, free ribosomes are present. Mitochondria are absent but respiratory enzymes are located on the surface of mesosomes which are invaginations of plasma membrane.

Golgi complex and true plastids are absent. However, photosynthetic bacteria show chromatophores containing bacteriochlorophyll (e.g. purple-S-bacteria) or chlorobium chlorophyll (e.g. green-S-bacteria). All three types of RNA are present. Cells are prokaryotic, i.e. without nuclear membrane, nucleoplasm and nucleolus. They contain a single molecule of circular double stranded DNA attached to the plasma membrane at one point.

Nutrition

They show autotrophic or heterotrophic mode of nutrition. Autotrophic bacteria are of two types, viz., photosynthetic, which produce food material by using light energy (e.g. purple-S-bacteria and green-S-bacteria), and chemosynthetic which produce food material by using chemical energy which is derived by oxidizing certain inorganic compounds (e.g. Nitrosomonas, Nitrobacter). Heterotrophic bacteria are saprophytic, growing on dead and decaying organic matter, or parasitic, growing in living plants and animals, including humans.

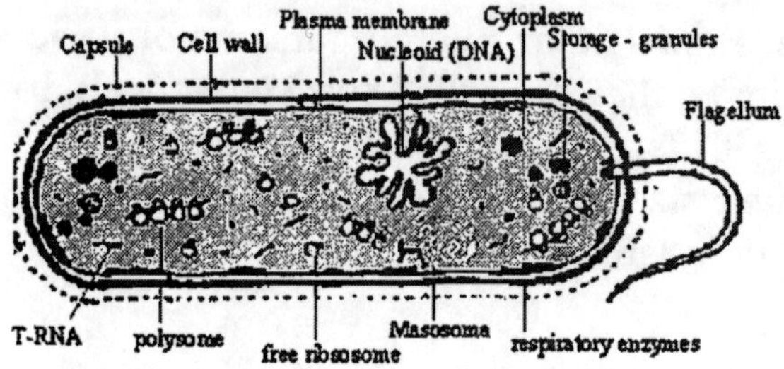

The parasitic bacteria may be pathogenic, causing diseases (e.g. Xanthomonas citri, Diplococcus pneumonae) or non-pathogenic, which usually form a symbiotic association with their hosts (e.g. Rhizobium, fixing nitrogen in leguminous plants, and cellulose-digesting bacteria in ruminant stomachs).

Staining

Those bacteria which retain Gram stain are called gram-positive bacteria, (e.g. Streptococcus) while those which do not retain the stain are called gram-negative bacteria, (e.g. Escherichia coli).

Reproduction

They reproduce both asexually as well as sexually. Asexual reproduction is by binary fission which takes place in favourable conditions, or by endospore formation, which takes place in unfavourable conditions. Sexual reproduction is by conjugation.

Economic importance

Bacteria act as both friends and foes of mankind as described below:

Useful activities

- They increase the fertility of soil by converting

atmospheric nitrogen into nitrates (nitrogen fixation), by converting ammonia derived from dead bodies into nitrates (nitrification) or by decomposing dead organic matter and forming humus which is the most fertile part of the soil.

- They form symbolic association with man and animals and help in digestion.
- In industry they are used in 'curdling' of milk, 'retting of fibers' of jute and hemp'curing' of tea and tobacco leaves 'tanning' of leather, production of vinegar, etc.
- They are used in the production of certain antibiotic drugs.
- They are scavengers of nature and help to keep the environment clean and also play an important role in recycling of nutrients in nature.

Harmful activities

- They cause dreadful diseases in plants, animals and human.
- They spoil milk, meat, fish and vegetables.
- Some of them release toxins into spoiled food and cause food poisoning.
- Denitrifying bacteria reduce fertility of soil by converting nitrates into free nitrogen (denitrification).

ACTINOMYCETES

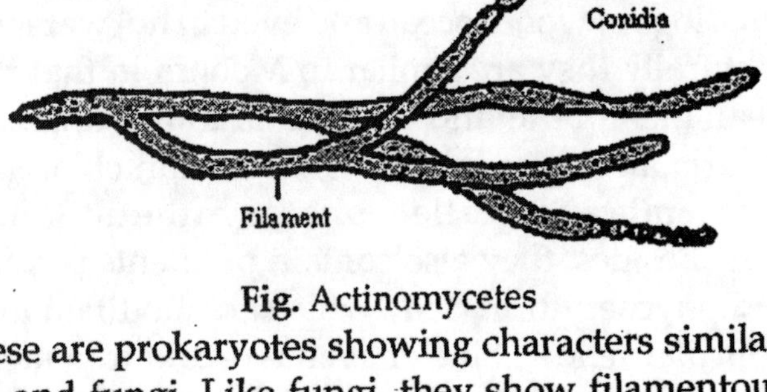

Fig. Actinomycetes

These are prokaryotes showing characters similar to both bacteria and fungi. Like fungi, they show filamentous bodies and they BIO reproduce by forming conidia in chains.

However, like bacteria, their Cell wall is made up of peptidoglycan and not of chitin or fungal cellulose. The nucleus is without a nuclear membrane, and sexual reproduction is totally absent. They are regarded as intermediate between fungi and bacteria; but because of the thin filamentous structure, they are classified with bacteria. They may be aerobic or anaerobic, saprophytic or parasitic.

The body of actinomycetes is made up of very thin, branched filamentous structure without cross septa. The branches grow in all directions. Some of the branches are aerial and produce conidia in chain at the apices. The cell wall is made up of peptidoglycan and the protoplasm shows granules and vacuoles. They are gram positive.

Economic Importance

Most of them are saprophytic, economically important as decomposers. Several species of the group Streptomyces produce antibiotics like streptomycin, erythromycin, tetracycline, neomycin etc while some are pathogenic like Mycobacterium tuberculosis causing tuberculosis in human beings, M. leprae causing leprosy, and others causing serious lung diseases.

Cyanobacteria

These were formerly known as blue green algae because of their filamentous nature. They are photosynthetic autotrophs widely distributed in nature. They are found in ponds, lakes, moist soils, logs of wood, ocean and even in hot water springs.

Structurally they are similar to Monera in that they have no nuclear membrane and no membranous organelles like mitochondria and chloroplasts. They contain chlorophyll and internal membranes called photosynthetic lamellae or thylakoids. Besides, they also contain pigments phycocyanine (blue) and phycoerythrin (red), which are modified to produce brown, purple, yellow, blue or even red-coloured individuals.

They are unicellular or colonial forms which are round, rod-shaped or filamentous. The cell wall is made up of peptidoglycan covered by a characteristic gelatinous envelope. Some cells in the filament are colourless and are called heterocysts which fix free atmospheric nitrogen. The fragmentations take place at these points.

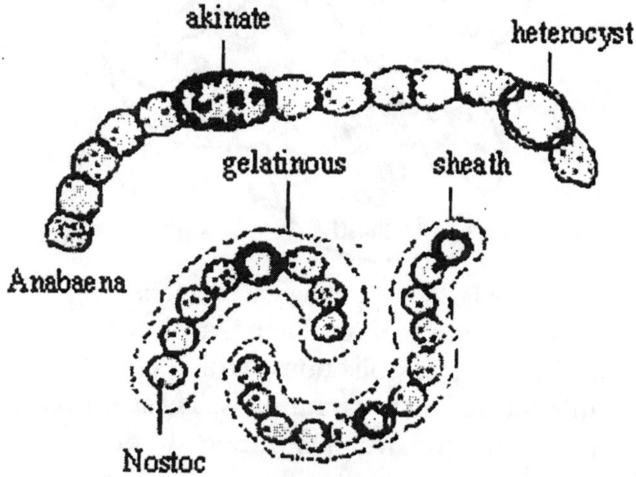

Fig. Cyanobacteria

They are devoid of flagella but some move by peculiar gliding movement. Food is often stored in the form of oil or glycogen droplets. Some can stand extremes of temperature and pH. The method of reproduction is by binary fission. Since they reproduce extensively, they often become dominant microorganisms in the polluted water of lakes and ponds containing a rich supply of organic matter, forming huge populations often called "blooms". The periodic redness in the oceans is due to blooms of red cyanobacteria. Common examples are Anabaena, Nostoc, Oscillatoria, etc.

Economic importance

Cyanobacteria have the following economic importance:

- Heterocyst-bearing forms perform the function of nitrogen fixation. Hence, they are used as biofertilizers in paddy field, to increase yield.
- Non-toxic forms like Spirulina are cultured in large tanks as protein-rich animal feed.
- Nostoc is cultured and used as feed for aquatic animals.

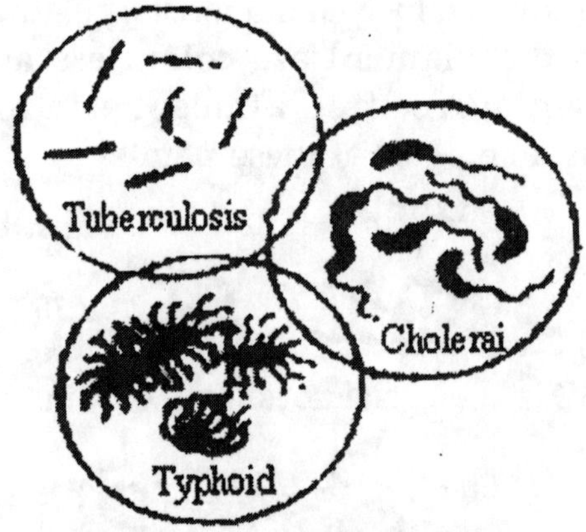

Fig. A few Pathogenic Bacteria

- Some forms like Anabaena and Nostoc make the drinking water poisonous, which can cause death of cattle, birds and even humans. These forms may even into erfere with water filtration systems.

Role of Monera

Role in cycling

Monera are chiefly decomposers concerned in cycling, rotation of metabolites like carbon, hydrogen, nitrogen, and sulphur. in nature. Life on earth would have run out were it not for the decomposition of dead matter by bacteria and release of the elements for resynthesis of cellular compounds. This cyclic journey of chemical elements from biological organisms (bio) through soil or earth crust (geo), is referred to as the biogeochemical cycle.

LIST OF COMMON BACTERIAL DISEASES IN MAN

No.	Disease	Causative agents
1	Dysentery	Shigella dysenteriae
2	Cholera	Vibrio cholerae
3	Diphtheria	Corynebacterium diphtheriae
4	Pneumonia	Diplococcus pneumoniae
5	Tuberculosis	Mycobacterium tuberculosis
6	Tetanus	Clostridium tetani

Role in Symbiosis

Some forms show symbiotic relationship such as Rhizobium in the roots of leguminous plants or Nostoc and Anabaena in coralloid roots of Cycas, which fix free atmospheric nitrogen into nitrates, and Escherichia coli which inhabit the colon of the human intestine and help in synthesizing Vitamin B.

Pathogenic Forms

Some Monera are also pathogenic, producing common diseases like typhoid, cholera, diphtheria, tuberculosis and pneumonia in human beings.

BROAD CONTROL MEASURES:

The broad control measures are as follows:

- *Use of disinfectants:* These are strong chemicals which are distributed where bacteria thrive and multiply. Some common disinfectants used are bleach, phenol, and H_2O_2 peroxide.
- *Use of antiseptics:* These are mild chemicals which are locally applied to kill bacteria, such as alcohol and iodine.
- *Use of antibiotics:* These are specific drugs which are used to prevent the growth of bacteria. Some common antibiotics are tetracyclin, streptomycin, penicillin, neomycin and erythromycin.
- *Sterilization:* This is the common method by which bacteria are killed by excessive heat or ultraviolet irradiation, which disrupts and destroys the protiens and nucleic acids in the bacteria.

HOT WATER SPRINGS

Hot water springs also called thermal springs are flows of hot water originating from active volcanic rock. The temperature of hot water is about 85° C. This water contains minerals (bicarbonates of calcium and magnesium) dissolved from the rock. At this temperature only certain bacteria like Thermus, Sulpholobus, Bacillus and some cyanobacteria can survive. Hence, these organisms are called thermophiles. Many times hot water springs contain sulfur. Hence, taking bath in such water is remedy for skin diseases in humans.

KINGDOM: PROTISTA

Eukaryotic unicellular organisms

Protista includes all eukaryotic unicellular microorganisms, either plant-like or animal-like or showing overlapping characters of both plants and animals. Primarily they are aquatic and widely distributed all over the world, occurring in oceans, lakes, ponds and damp soils. They are autotrophic or heterotrophic.

The latter are free-living or parasitic on or within multicellular organisms. Thus it reflects the lifestyles either of plants, animals or fungi. Phylogenetically they serve as the connecting link between prokaryotic Monera and complex multicellular kingdoms of plants fungi and animals. They are distinguished by the following characters:

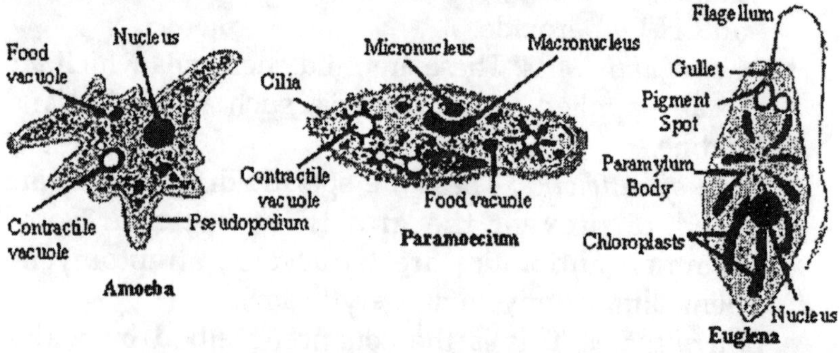

Fig. Some Protista

- They are first eukaryotes, having a well organized nucleus and complex membranous organelles.
- They are unicellular or colonial forms without distinct division of lobour.
- They are autotrophic or heterotrophic showing varieties of metabolic systems.
- Locomotion is by pseudopodia, flagella or cilia.
- They show mitosis, meiosis and simplest type of sexual reproduction for the first time. Common examples are Ameba, Paramecium, Euglena, diatoms and dinoflagellates.

CILIA AND FLAGELLA

These are microscopic, contractile, motile hair-like locomotary organelles present in ciliated and flagellated protists like Paramecium and Euglena. Cilia and flagella are similar in structure. Each arises internally from the basal body and is made up of eleven microtubules, two single ones in the centre connected to nine double ones arranged along the periphery (often referred to as 9 + 2 arrangement). The outer fibrils enter cytoplasm and converge to form the basal body. The central fibrils are connected to the peripheral ones by radial lamellae, like the spokes of a wheel.

Cilia are short and numerous and beat in a coordinated manner simultaneously or one after the other, while flagella are long and whip-like, showing undulating movements. Their function is to propel the cell through the surrounding liquid medium or move the surrounding medium past the cells, gathering food particles. During movements cilia beat vigourously and rapidly (effective stroke) and they recover slowly (recovery stroke). Cilia and flagella are widely distributed in gametes, unicellular plants and animals and also on cells of more complex organisms forming the internal lining of ducts, such as trachea, oviduct, etc.

All living organisms start their life with a single cell (zygote). In unicellular animals, cell division means reproduction which results in the formation of new organisms. In multicellular organisms, however, this cell further divides and thus forms the basis of growth. What initiates cell division is not exactly known but it is believed that when the ratio between the volume of cytoplasm and nucleus, (called nucleoplasmic index) exceeds its limit, the cell divides. Principally, there are two types of cell divisions:

- Mitosis which is characteristic division of somatic or body cells.
- Meiosis which is characteristic division in which germ cells are formed in the reproductive organs.

Cell Cycle and Mitosis

The period from the beginning of one cell division to the beginning of the next cell division or the period between the end of one cell division to the end of next cell division, is known as cell cycle. It includes two main phases, that is, interphase and mitotic phase.

The interphase is divided into following three phases.

- First gap stage or G_1 phase in which cell grows.
- S phase or synthetic phase in which DNA replicates.
- Second gap phase or G_2 phase in which the protein synthesis increases. The completion of G_2 phase is marked by the beginning of mitosis.

The role of mitosis in cell cycle is to form two sets on identical chromosomes and to distribute one complete set at each end to form a daughter nucleus. Mitosis thus ensures that the new cell contains the identical number and types of chromosomes present in original mother cell, thus to continue the life-cycle.

In unicellular animals, reproduction is by mitotic division but in higher organisms with evolution of sex, reproduction takes place by the union of two gametes – male and female (sperm and egg). Hence, it is essential that at some stage (gamete formation) during life cycle the number of chromosomes should be reduced to half, so that their union will maintain the original number of chromosomes constant. This is done by meiosis.

MEIOSIS

Meiosis is the characteristic division of the germ cells. It is also called reduction division because in this the parent cell produces four daughter cells (tetrads) each having half the number of chromosomes present in the parent cell. In other words, meiosis is a process of cell division in which diploid (2n) number of chromosomes is reduced to haploid (n) number. Meiosis is completed in two distinct nuclear divisions which take place in quick succession, and are respectively called:

- First meiotic or reductional division in which the number of chromosomes is reduced to half.
- Second meiotic division or equational division in which the number of chromosomes is the same as at

the end of first division. Each division of meiosis like mitosis consists of four stages prophase, metaphase, anaphase and telophase.

Cell division involves two stages, the nuclear division called karyokinesis and the division of cytoplasm known as cytokinesis. In any cell division the nucleus divides first followed immediately by the division of cytoplasm. It is the characteristic division of somatic or body cells resulting in the formation of daughter cells each receiving equal number of chromosomes like those of parent cells. Hence it is called equational division. It can be described into four orderly and consecutive phases such as prophase, metaphase, anaphase and telophase.

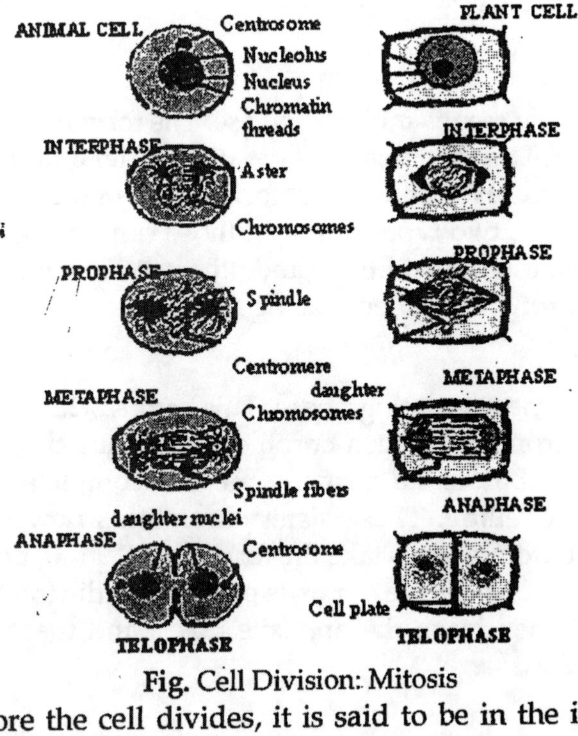

Fig. Cell Division: Mitosis

Before the cell divides, it is said to be in the interphase which is sometimes inappropriately called 'resting phase', though the cell is definitely not resting; in fact, all its metabolic or synthesizing activities are at its peak as the cell has to prepare itself for division. However, it is characterized by a large and distinct nucleus but indistinct chromosomes.

Prophase

It literally means preparatory phase. It is initiated by the changes in the centrosphere. The centrosphere disappears and the two centrioles that are set free begin to move away from each other. Each of them radiates fibers, called an aster. The asters are absent in plant cells. The long chromatin threads of DNA undergo condensation to form a definite number of distinct chromosomes by dehydration and coiling. Each chromosome is double, consisting of two identical components called the chromatids connected by a common body called centromere. The nucleolus and nuclear membrane begin to disappear and the fibers radiating from pole to pole, called spindle fibers begin to appear.

Metaphase

It literally means mid-way phase. The formation of spindle is complete. The chromosomes get still shorter and thicker and arrange themselves on the equatorial plane of the spindle. The spindle shows two types of fibers, those running from pole to pole (called interzonal fibers) and others connecting each pole to the centromere, (called tractile fibers).

Anaphase

This is the separating phase. This begins with the division of the centromere of each chromosome. Thus the two sister chromatids of the same chromosome now completely separate and start repelling. These sister chromatids now known as daughter chromosomes take the form of a V or a U, and appear to be pulled towards the opposite poles along the tractile fibers. Gradually they reach the opposite poles, and that marks the end of anaphase.

Telophase

This is the reconstruction phase. It includes the changes that are reversal of those in prophase. The chromosomes aggregate at the opposite poles and once again get elongated, uncoil and become indistinct. The spindle fibers simultaneously begin to disappear and the two daughter nuclei are formed at opposite poles, each containing an equal number of chromosomes as those of the parent nucleus.

Cytokinesis

It usually starts with the beginning of telophase. In animal cells it proceeds by a cleavage furrow around the cell at the equatorial region. This furrow increases and deepens until the cell is divided into two. Mitosis in animal cell and plant cells are basically similar except that in plant cells there are neither centrioles nor asters. Moreover, cytokinesis in plant cell is not accomplished by furrowing but by formation of a cell plate between the seperate sets of chromosomes. The cell plate is impregnated by pectin to form middle lamella. Later on cellulose is laid down on either side of the middle lamella to form the cell wall, thus dividing the mother cell into twodaughter cells.

- Daughter cells receive same number of chromosomes as in parent cell. Thus it helps to maintain genetic integrity of the species.

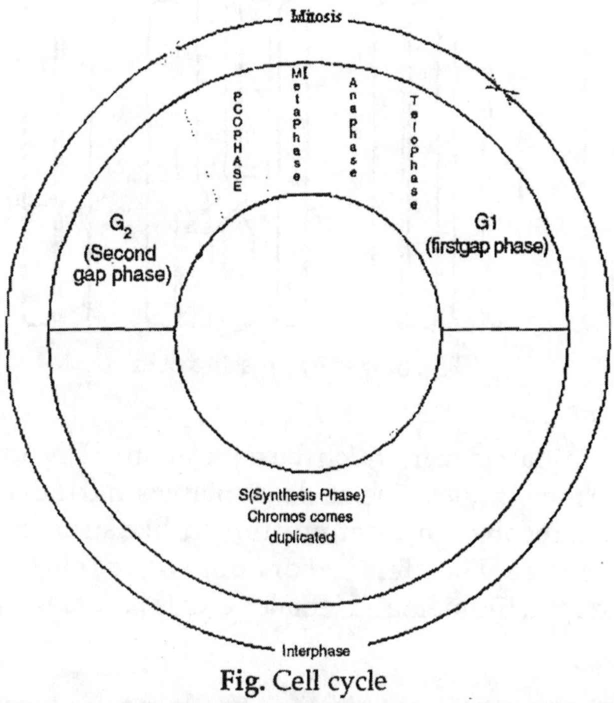

Fig. Cell cycle

- Since equal number of chromosomes are distributed to

the daughter cells, the daughter cells are similar to the parent cell quantitatively as well as qualitatively.

- It is responsible for the replacement of old cells like those of the epidermis and blood corpuscles, having a short span of life.
- It promotes growth development of the organism.

In multicellular organisms, cell division either contributes to the cell replacement from wear and tear of the body (regeneration) or adds to the cell number, which results in the growth of the organism. First meiotic division Prior to meiosis some changes take place in the nucleus in which DNA begins to replicate. It is sometimes known as interphase.

Prophase

The prophase of meiosis 1 is much prolonged and can be divided further into five stages as follows:

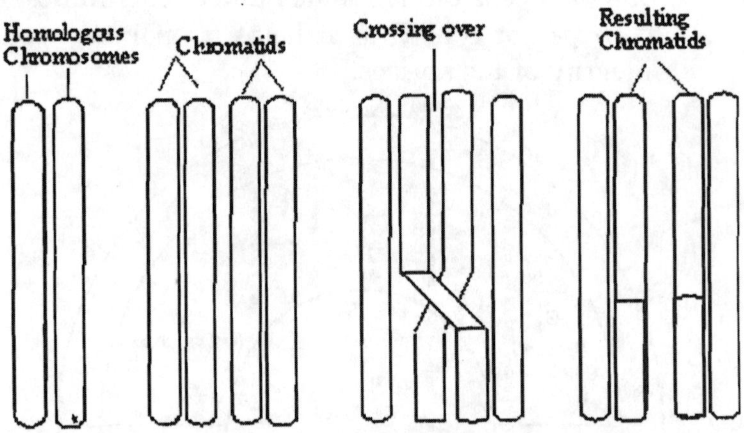

Fig. Stages in Crossing over

Leptotene

The initial stage in which the replication of DNA started in interphase continues. The nucleus enlarges and the chromatin reticulum resolves into definite thread-like structures called chromosomes. The identical or homologous chromosomes received from the male and female parents become discernible.

Zygotene

In this stage the homologous chromosomes begin to pair

length to length and point to point in a zipper- like manner. This pairing is called synapsis, and pairs are referred to as bivalents formerly known as dyads (synapsis is absent in mitosis).

Pachytene

The paired chromosomes at this stage become shorter and thicker. The chromosomes of each bivalent then split lengthwise to form four chromatids (tetrads).

Diplotene

The force of attraction between the bivalents lapses;, however, they remain attached at certain points called chiasmata where mutual exchange of parts, called crossing over, can take place.

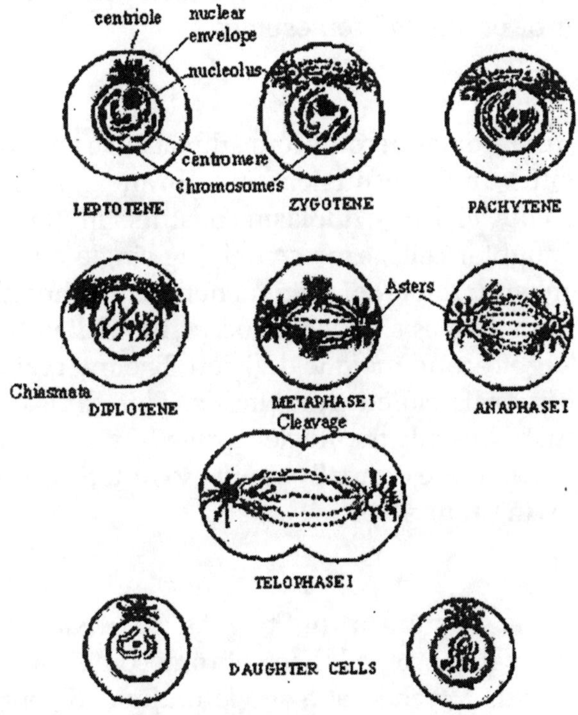

Fig. Cell Division: Meiosis I

Diakinesis

The shortening of chromosomes reaches its maximum. They assume characteristic forms depending upon the position of chiasmata. The nucleolus and nuclear membrane begin to disappear and spindle begins to appear.

Metaphase I

The spindle formation is complete. The bivalents move and arrange themselves on equatorial plane. Each chromosome of the bivalents gets attached to the spindle fiber at the centromere.

Anaphase I

The two chromosomes, or mates, of a bivalent (not the chromatids of the same chromosome) start moving towards the opposite poles. Thus the real reduction in the number of chromosomes takes place at this stage, each pole receiving half the number of original chromosomes.

Telophase I

The chromosomes become long, thread-like and indistinct. The nucleolus and the nuclear membrane reappear thus forming two nuclei. The cytoplasm constricts in the middle to form two daughter cells, each containing single nucleus with half the number (haploid) of original number of chromosomes.

Usually cytokinesis does not occur immediately but the nuclei undergo second meiotic division. The interval between the two division is called interkinesis. This division occurs simultaneously in both the nuclei formed as a result of first meiotic division. It is essentially similar to mitotic division and shows following four steps:

Prophase II

The nuclear membrane begins to disappear and chromosomes get shortened. Each chromosome showing two sister chromatids attached at a single undivided centromere.

Metaphase II

The spindle is formed usually at right angle to the plane of the spindle formed in Meiosis I. The chromosomes get arranged in an equatorial plane with centromeres attached to spindle fibers.

Anaphase II

The centromere of each chromosome now divides for the first time separating sister chromatids which travel to the opposite poles.

Telophase II

The set of chromatids at each pole organizes into a new daughter nucleus with the reappearance of nuclear membrane and nucleolus. Thus four daughter nuclei are formed each containing haploid number of chromosomes. Cytokinesis follows and four daughter cells are formed (tetrads) at the end of Meiosis-II, each with the nucleus containing haploid number of chromosomes. Meiosis, therefore, is a profound modification of mitosis in which the chromosomes first pair and then divide to form four haploid cells.

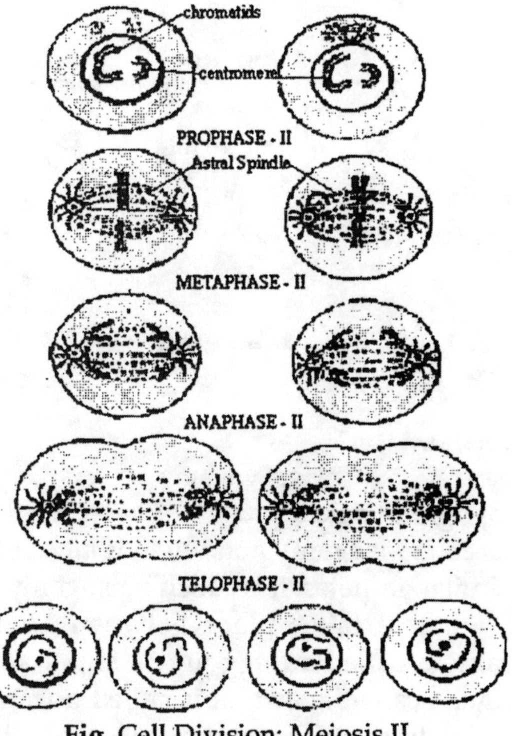

Fig. Cell Division: Meiosis II.

Significance of Meiosis

- It compensates for the doubling of chromosomes at

gametic fusion and thus helps to keep the number of chromosomes constant, characteristic of each species.

- It is responsible for giving rise to large number of variations in the offsprings by crossing over and random distribution of maternal and paternal chromosomes.
- The variations coupled with heredity form the raw material for organic evolution. If it was not for meiosis, the evolution of species would be suspended.

SYNGAMY

Syngamy is the fusion of two haploid nuclei or gametes.

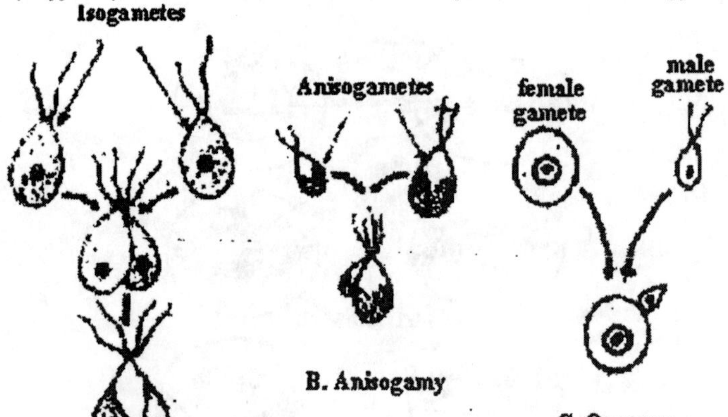

Fig. Syngamy

When syngamy takes place between the two identical gametes, it is called isogamy and when it takes place between the two dissimilar gametes, it is called anisogamy. As in higher organisms, when it takes place between a large, passive receptive gamete called an egg and a small active (motile) initiative gamete called sperm, it is called oogamy.

Meiosis leads to the formation of haploid sex cells or gametes which are individually incapable of further development. Thus, the sex cells or gametes are arrested in their growth, but the syngamy removes the arrest and the diploid condition is restored. The diploid zygote formed by the union of haploid gametes is potentially capable of developing into a new individual.

EVOLUTIONARY PRECURSORS OF COMPLEX LIFE FORMS

Protista is a diversified assembly of different groups of organisms which includes:

- Unicellular algae-like protista, e.g. Diatoms, Dinoflagellates etc.
- Fungi-like protista, e.g. slime molds
- Animal-like protista, e.g. protozoa

DIATOMS

These are microscopic, usually single-celled or colonial forms. They are found both in fresh water and salt water. In ocean they occur in vast assembly as floating plankton. They show a variety of patterns showing extreme beauty and symmetry because of which they are sometimes called "jewels of the plant world."

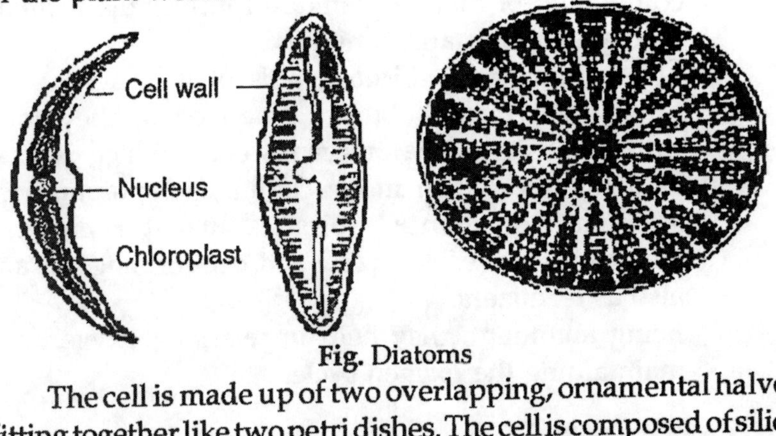

Fig. Diatoms

The cell is made up of two overlapping, ornamental halves fitting together like two petri dishes. The cell is composed of silica and manganese instead of cellulose. It is uninucleated with ribbon-like plastids, containing chlorphyll-a, chlorophyll-c, beta-carotene and oxanthophyll. They store food as oil and polysaccharides other than starch. The locomotion is of gliding type produced by the streaming of cytoplasm through raphae or grooves on the surface of the cell wall. There are two types of forms, viz. pinnate with bilateral symmetry, and centric with radial symmetry.

Common life-cycle

Diatoms are diploid. They reproduce asexually as well as sexually. In asexual reproduction it divides by mitosis in which two valves are separated, each cell receiving one valve of the parent cell. Each of the two cells forms a new valve fitting into the old one. Thus, the size goes on reducing during repeated mitotic division.

At a certain stage the protoplast escapes to form two haploid gametes by meiosis. These gametes fuse to form diploid zygotes called auxospores which increase in size to compensate for the reduction, before the new cell wall is laid down. Thus, the life-cycle shows the alternation of diploid and haploid generations.

Common Examples are *Pinnularia and Navicula.*

Economic importance

- The diatoms play an extremely important role in the aquatic food web. They are the most abundant component of marine plankton forming the primary food source of marine animals.
- The imperishable siliceous cells of diatoms are led down in various habitats. When dead, they form thick deposits of diatomaceous earth which is used as an ingredient in many commercial preparations such as detergents, polishes, paint removers, insulators, for decolourizing and deodorizing oils and also as fertilizers.
- Being abundant, they help in releasing oxygen and maintaining the oxygen cycle.

Diatomaceous Earth

The imperishable, siliceous dead cells of diatoms do not disintegrate but form thick deposits called diatomaceous earth or diatomite. It is procured from exposed deposits and lake bottoms. The largest deposits of diatomaceous earth are found in California. In Lampoc it is extending miles long and with average depth of 425 meters while in the oil fields of Santa Maria, it is 3000 feet thickness.

DINOFLAGELLATES

These are one celled, biflagellate organisms, mostly marine and are important photosynthesizers. Some species are bioluminiscent, often seen in ocean waters at nights. They are commonly called fire algae as they emit red light. Their cell wall is made up of cellulose in the form of interlocking cellulose plate-forming theca. Most species have two unequal flagella hence called dinoflagellates. One flagellum runs lengthwise and extends behind like a tail and another encircles the body like a belt, in a transverse groove. The rotating or spinning movements of dinoflagellates are due to these flagella. The photosynthetic species have discoid plastids containing yellowish green carotenoids, chlorophyll and xanthophyll.

They store food in the form of oil as well as polysaccharides. The nucleus is unique in that it has chromosomes permanently condensed and remain attached to nuclear membrane. Mitosis is intranuclear, i.e. it takes place within the nuclear membrane. They differ from diatoms in having flagella, absence ot glossy skeleton of silicon and the cell wall is made up of cellulose plates.

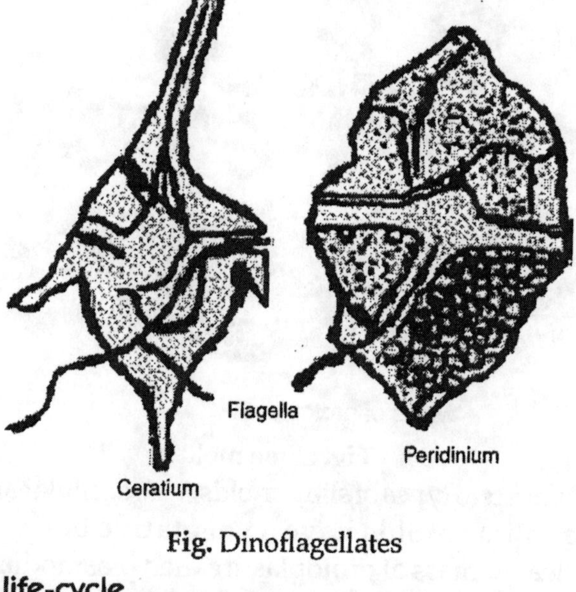

Fig. Dinoflagellates

Common life-cycle

Asexual reproduction takes place by simple cell division

(fission) or by zoospore formation. Recently sexual reproduction by isogamy has been reported in some. The examples are Ceratium, Noctiluca, Peridinium, Ganyaulux. Economic importance: They form an important constituent of marine plankton and thus play an extremely important role in the aquatic food web. A number of dinoflagellates like Ganyaulux are toxic, forming red tides killing a number of fish. Some species cause dermatitis in swimmers. They reproduce both sexually and asexually and the life history shows alternation of generations as shown by diatoms.

Slime molds

General account: Slime mold is a mass of naked protoplasm without a rigid cell wall. It creeps along the substratum by ameboid movements and engulfs spores, bacteria and particulate debris of the substratum on which it grows. It commonly reproduces by structures called fruiting bodies which produce spores.

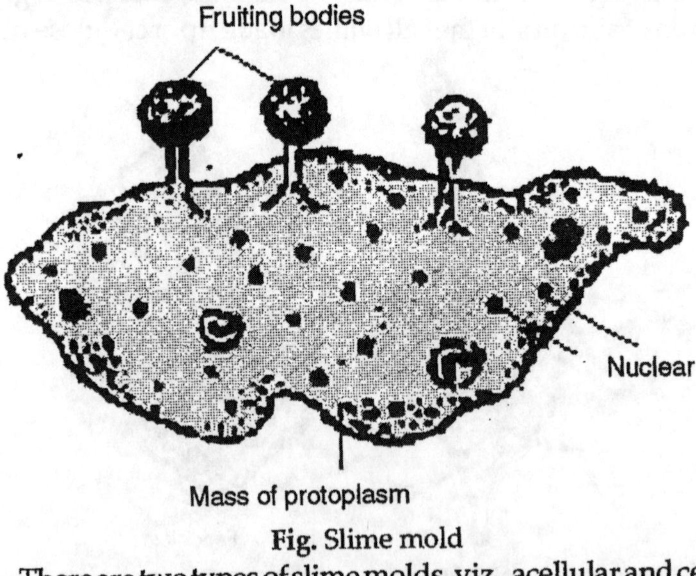

Fig. Slime mold

There are two types of slime molds, viz., acellular and cellular. Acellular slime molds have a vegetative body which is a multinucleated mass of protoplasm called plasmodium. It starts its life as a single diploid cell showing ameboid movements. Soon the nucleus divides repeatedly forming a large mass of naked protoplasm with number of nuclei. Every now and then it changes its shape and grows. Under unfavourable conditions, it develops little mounds on plasmodium, each of which gives a stalk bearing a sporangium at the tip. The spores are formed by meiosis, hence they are haploid. Each spore produces one to four flagellate spores which behave like undifferentiated gametes and fuse in pairs to form diploid zygotes. The zygote divides repeatedly by mitosis to form new plasmodium.

Economic Importance

In recent years slime molds have become important tools of research to study structure, streaming of movements and the physiology of protoplasm.

PROTOZOANS

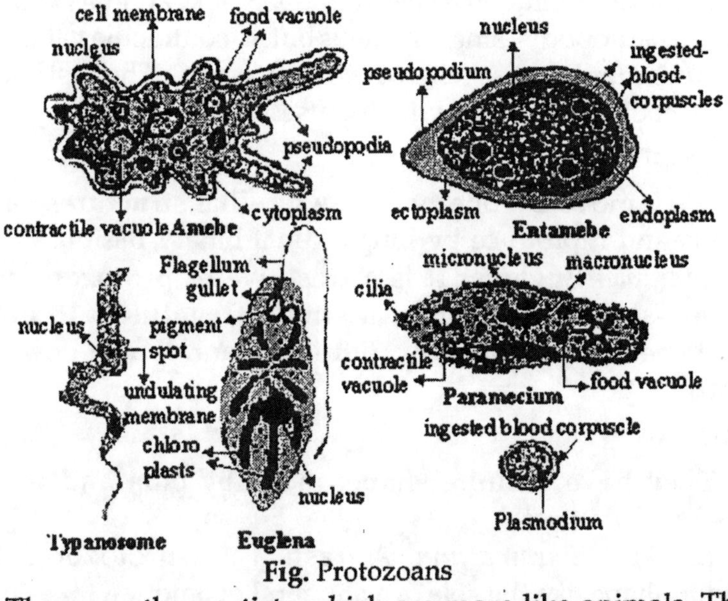

Fig. Protozoans

These are the protists which are more like animals. They are unicellular organisms found in fresh water, salt or any moist surroundings. They may be free living or parasitic, solitary or found in loose colonies.

They are distinguished by the following characteristics:

- The body is covered by plasma membrane or labile pellicle and contains one or many nuclei.

- They are generally motile. Locomotion is carried out by flagella, cilia or pseudopodia.
- Most of them are holotrophic though some are saprophytic and others parasitic in nutrition.
- They show different organelles (cell organs) like food vacuoles, contractile vacuoles and, in some, cytostome and cytopharynx to perform various functions of life.
- They are sensitive to various external stimuli like touch, temperature, light etc.

The protozoans fall into following four groups:

Sarcodina

They have no definite shape, move by pseudopodia and reproduce by binary fission.

Example: Entamoeba: These are ameba-like animals with one or two pseudopodia, single nucleus but no contractile vacuole. It includes a number of species which are parasitic. The best known is Entamoeba histolytica, which causes amebic dysentery.

Mastigophora or flagellata

They move by one or two whip-like structures called flagella and reproduce by longitudinal binary fission.

Example: Euglena: It is a fresh water protozoon with spindle-shaped body and a single flagellum. It shows cytostome and cytopharynx, but like plant, it also possesses chloroplasts.

Ciliophora

They have definite shape, move by cilia and mostly possess two nuclei.

Example: Paramecium: A fresh water protozoon with definite shape like the sole of a slipper. Locomotion takes place by cilia. The animal has usually two nuclei and two contractile vacuoles. It reproduces by simple type of sexual reproduction called conjugation.

Sporozoa

These are parasitic forms. The locomotory organelles, food vacuoles and contractile vacuoles are absent.

Example: Plasmodium: An intracellular blood parasite causing malaria. The life cycle is completed in two hosts, namely man and mosquito.

Economic Importance: Most of the protozoa are harmless but there are a few parasitic forms, such as Plasmodium, which causes malarial fever, Trypanosoma which develops sleeping sickness, and Entamoeba histolytica which causes dysentery. Protozoa are useful in sewage disposal.

SYMBIOTIC FORMS OF PROTISTA

These are the organisms of different species which live in intimate relationship for mutual benefit. Examples are lichen, and Trichonympha in the intestine of termites.

LICHEN

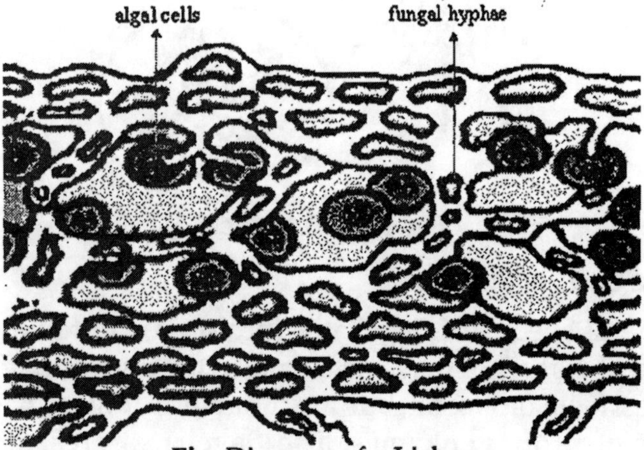

Fig. Diagram of a Lichen

It is an obligatory association between, and plant, which together form a closely integrated unit called a lichen. The body of the lichen is composed of branching hyphae of the fungus which harbor algal cells. The fungus gets food synthesized by the alga while the alga in return gets shelter, moisture and minerals from the fungal partner.

Economic Importance

Lichens are used as food by human beings. They are also used in medicine, as dye stuffs for tanning hide into leather, etc. some like Usnea, are responsible for skin diseases, respiratory allergy etc.

Root Tubercles

An obligatory association between plants and microbes, i.e. leguminous plant and the bacteria called Rhizobium. The latter are nitrogen fixing bacteria living in the root modules of leguminous plants like pea, bean, etc. Rhizobium fixes free nitrogen in the form of nitrates and makes it available to the plant and in return gets food and shelter from the latter.

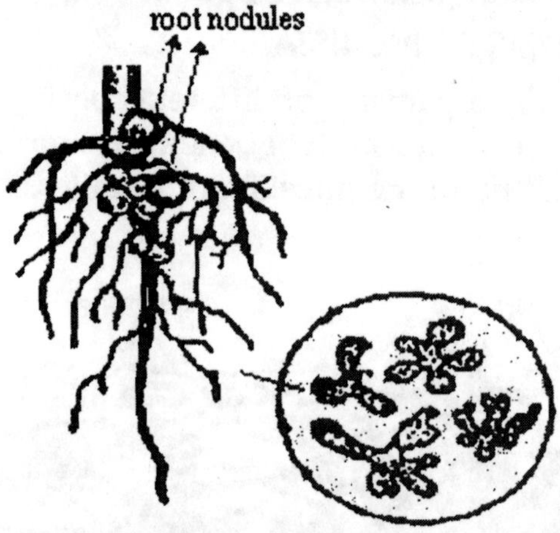

Fig. Root Tubercles

Termites and Flagellates

Several species of flagellates (protozoa) are found in the gut of termites where they participate in the digestion of wood. Termites feed on wood but are unable to digest it due to absence of proper enzymes. The flagellates present in the intestine digest wood with specific enzymes they secrete. The flagellates have no mouth or gullet; food particles are engulfed by pseudopodia at the posterior end. They convert the cellulose of wood into sugars. They use some of the sugars but much is left for the termites. The termites cannot survive without their intestinal flagellates. Newly hatched termites lick the anus of other termites to obtain flagellates. The flagellates in return get protection, food and lodging. Thus, the flagellate and termite live together for mutual benefit.

Ruminant Stomach

Cud chewing animals like cow, sheep, goat, deer, giraffe

etc. are called ruminants. They are herbivores. In these animals while grazing, food is cut into large pieces and swallowed without mastication. When at rest, the cud (i.e. the food once swallowed) is regurgitated into the mouth for thorough mastication, a process called rumination, or chewing the cud.

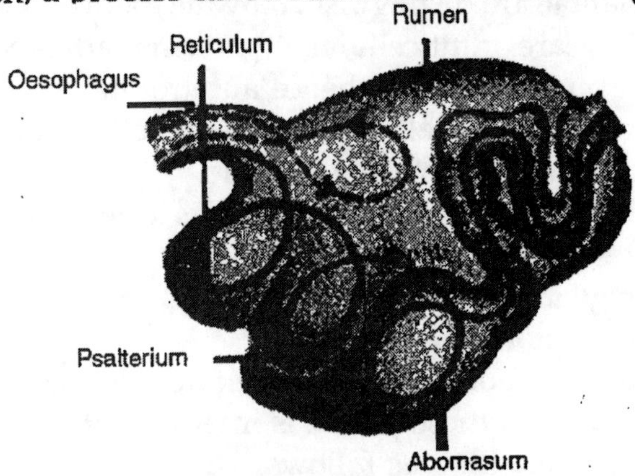

Fig. Ruminant stomach

In adaptation to this peculiar food eating habit the stomach of a ruminant is very modified. It consists of four interconnected chambers such as:

- Rumen or pouch
- Reticulum or honeycomb
- Psalterium or manyplies or omasum
- Abomasum or runnet.

The rumen is the first chamber in which food is received without mastication. The rumen is much enlarged and serves not only as a storehouse of food that is swallowed without mastication but also acts as a fermentation chamber due to its high contents of microbes. These symbiotic microbes digest cellulose which the ruminant cannot hydrolyze due to lack of cellulase. They also digest lipids into fatty acids and synthesize vitamin B. Absorption of these products by the rumen provides a significant amount of energy to the animal.

KINGDOM: PLANTAE

Kingdom Plantae includes all organisms which are truly multicellular and photosynthetic exept for algal protists. They are complex autotrophs preparing food material by photosynthesis for themselves as well as for rest of the other organisms. Hence, they are the chief producers of the world. They are found in all the types of environment: aquatic algae, amphibian mosses, and terrestrial ferns and seed-bearing plants.

The plantae are distinguished by the following characters:

- They are multicellular organisms, adapted to carry on photosynthesis, hence autotrophic.
- Presence of cell wall of true cellulose, enclosing cytoplasm with large vacuole.
- They perform photosynthesis due to chlorophyll present in chloroplasts.
- They may show alternation of sexual and asexual generations in life-cycle.
- They are non-motile, without definite size and shape.
- Growth in these plants is intermittent.

They are classified as follows:

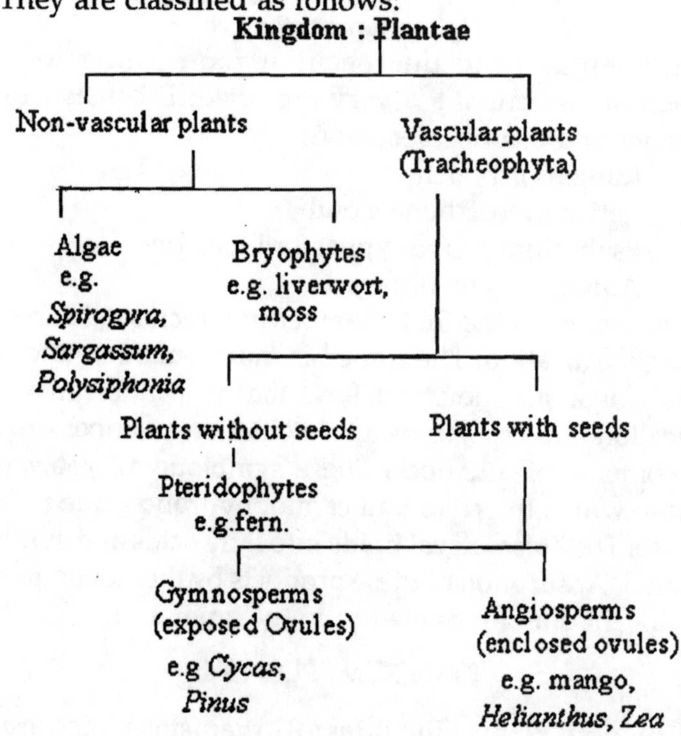

ALGAE

Algae are chlorophyll-bearing unicellular or multicellular

plants. When multicellular, they may be colonial or filamentous. Most of them are aquatic, either fresh water, (Volvox), or marine, (Spirogyra). Some are sheet-like (e.g., Ulva). Chlorophyll is present in chloroplasts, the number and shape of which are characteristic of each alga. Besides chlorophyll, they also show various carotenoid pigments which impart different colours to algae. According to the nature of photosynthetic pigments, they are further classified into three divisions such as Chlorophyta (green),. Phaeophyta (brown), and Rhodophyta (red).

Life - cycles

Algae show both a gametophytic (haploid) stage and a sporophytic (diploid) stage, which alternate. The life-cycle mainly shows two types, either haplontic or diplontic.

- Haplontic life - cycle, when the plant body is a gametophyte i.e., haploid and is dominant or of long duration, and the zygote is diploid and of short duration.
- Diplontic life-cycle, when the plant body is a sporophyte, i.e. diploid and is dominant or of long duration, and the gametophyte is haploid and of short duration.

Green Algae (Chlorophyta)

These algae live in wide variety of habitats, marine to fresh water to damp soil.

- These are unicellular (Chlamydomonas), colonial (Volvox) or filamentous. When filamentous they are unbranched (Spirogyra, Ulothrix) or branched (Chara).
- The cell wall consists of an inner layer of cellulose and outer layer of pectic compounds and may be covered by a gelatinous sheath.
- The protoplasm is divisible into cytoplasm and nucleus. Cytoplasm contains one or more vacuoles. Chlorophyll is present in chloroplasts, the shape and number of which are characteristic of each alga.

- Pigments chlorophyll-a and chlorophyll-b are predominant. However, carotene and xanthophyll are also present.
- Food reserve is in the form of starch surrounding the proteinaceous refractile bodiës called pyrenoids.
- Reproduction is vegetative by mitotic cell division in unicellular forms or by fragmentation in filamentous forms; asexual by formation of spores such as zoospores, aplanospores, hypnospore and akinates and sexual simply by conjugation or by gamete formation by isogamy or anisogamy or oogamy.
- The life-cycle is of haplontic type showing alternation of dominant haploid stage with short-lived diploid stage.

Examples: Chlamydomonas, Volvox, Ulothrix, Spirogyra, Ulva etc.

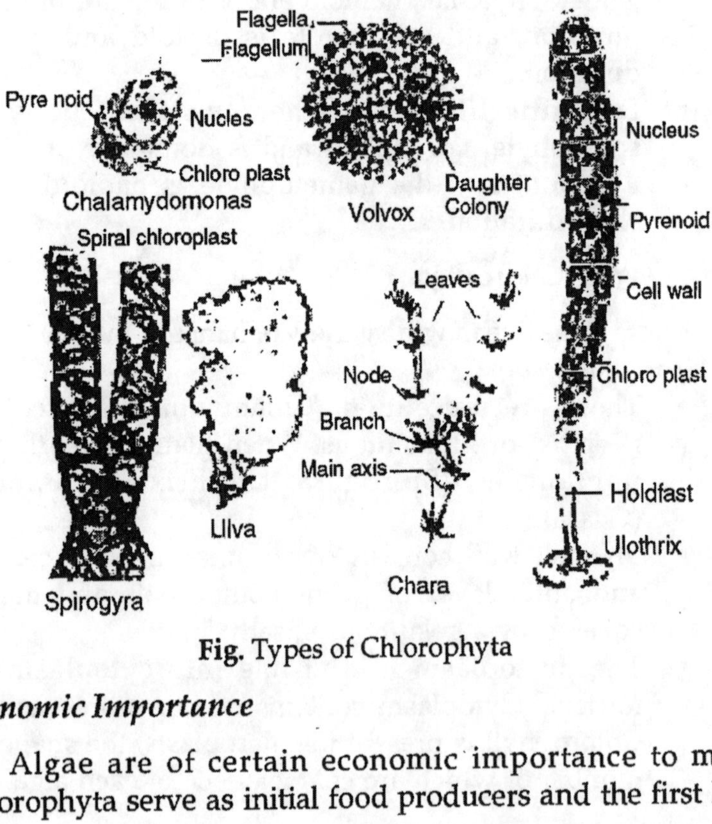

Fig. Types of Chlorophyta

Economic Importance

Algae are of certain economic importance to man Chlorophyta serve as initial food producers and the first link in the aquatic food chain, both fresh water and marine. Some like Ulva and Chlorella are used as vitaminized food.

Evolutionary Significance

It is believed that chlorophyta has evolved from some flagellate unicellular ancestors like Chlamydomonas. These ancestors, during evolution, gave rise to complex filamentous forms which supposed to have given rise to land plants like bryophytes.

Brown Algae (Phaeophyta)

These are multicellular, simple, filamentous or plant-like giant forms called kelps (sea weeds). These are mostly marine, found in cool shallow water.

- The body is the multicellular thallus showing the highest degree of differentiation. The unicellular colonial forms are absent. Some brown algae called kelps or sea weeds exhibit giant forms extending over 50 meters in length and showing parenchymatous organization.
- The cell wall is made up of two layers. The inner firm layer is of cellulose while the outer layer is gelatinous containing compounds like algin, fucosin etc.
- The cytoplasm show one or many vacuoles and single large, distinct nucleus, with one or more nucleoli. Chloroplasts are either discoidal or band-shaped, without pyrenoids.
- Yellow-brown pigment fucoxanthin a type of xanthophyll is predominant which gives golden - brown colour to algae, while other pigments such as chlorophyll-a and chlorophyll-b are also present.
- The reserved food is in the form of soluble carbohydrates called laminarin and mannitol.
- Reproduction is vegetative by fragmentation, asexual by formation of haploid or diploid zoospores and sexual by gamete formation, by isogamy, anisogamy or oogamy.
- The life-cycle shows isomorphic alternation of generations where the gametophyte and sporophyte

are morphologically similar (e.g., Ectocarpus), heteromorphic alternation or generations where the gametophyte and sporophyte are morphologically differentiated (e.g. Laminaric) while in there is no alternation of generations as gametophytic generation represented only by gametes.

Examples: Fucus, Sargassum, Laminaria, Ectocarpus etc.

Economic Importance

"Komba", a product made from Laminaria, is widely used as food in Japan and oceanic islands. It is boiled with fish, meat or in soups or as a cooked vegetable. A gel called algin, obtained from Macrocystis, Laminaria, etc., is used in the preparation of toothpastes, ice-cream, etc. Sea weeds like Sargassum are used as a directory source of iodine and bromine.

Evolutionary Significance

The universal appearance of motile gametes probably suggests that Phaeophyta took its origin from unicellular brown, flagellated ancestors. These do not seem to be related to green algae because of the multicellular vegetative body and presence of fucoxanthin and laminarin which belong exclusively to this group.

Red Algae (Rhodophyta)

These are exclusively marine, found in deep waters attached to rocks.

- These are always multicellular, filamentous, radially symmetrical or compressed forms. There is no true parenchymatous construction.
- The cell wall is made up of two layers. The inner one is of cellulose while the outer one is of pectic compounds with mucilaginous envelope.
- The cytoplasm is uninucleate and shows a central vacuole. The chromatophores are with naked pyrenoids.
- The pigments are chlorophyll-a, chlorophyll-b, carotene and xanthophyll. However, the green colour of chlorophyll-a is masked by red pigment, phycoerythrin and blue pigment phycocyanin.

- Food is stored in the form of chemically distinctive starch called floridean starch.
- The reproduction is vegetative by fragmentation, asexual by non-flagellated, haploid carpospores, monospores or tetraspores, and sexual by formation of non-flagellate, non-motile male gametes and an egg, thus oogamous.
- The life-cycle shows either haploid forms or regular alternation of generations between similar haploid and diploid stages.

Examples: Agardhiella, Palysiphonia, Batrachospermum, Gracilaria, etc.

Economic Importance

The gelatinous substance extracted from red algae (Gelidium, Gracilaria) is used to prepare agar as a medium for bacterial and fungal cultures, in preparation of ice-cream, jelly, cosmetics, medicines, etc. in Japan. Agar is also used in printing and dying processes in textile industry. Some species (Porphyra) are used as food.

Evolutionary Significance

It appears that this group has taken its origin from some unicellular, non-ciliate ancestor as it shows non-flagellated spores and gametes.

Economic Importance of Algae

Algae are of certain economic importance to mankind. They serve as initial food producers and the first link in the aquatic food chain, both fresh water and marine. Some of the fresh water algae and sea-weeds are used as vitamin-rich food. Brown algae contain iodine and algin. Some red algae are the source of agar jelly, used in the preparation of ice creams and culture media. However, algae sometimes cause contamination of water and some of them release many toxic substances killing millions of fish and other animals drinking this water.

BRYOPHYTA

These are terrestrial non-vascular plants which still require moist environment to complete their life-cycle, hence they are called amphibians of the plant kingdom. Bryophytes show advances over alga by developing special sex organs like antheridia (male) and archegonia (female) and show distinct alternation of generations. Bryophyta are the simplest of land plants, and include mosses, liverworts and hornworts.

They are distinguished by the following characteristics:

- They are small terrestrial plants.
- They are without a distinct root system but attached to substratum by rhizoids.
- They do not posses true vascular tissue.
- Sex organs are multicellular with a protective jacket layer.
- Gametophyte is dominant and independent.
- Sporophyte is small and parasitic or semiparasitic on the gametophyte.
- They show a distinct alternation of generations.

Bryophytes are Divided into three Classes

Bryophyta

Hepaticeae	Anthocerotae	Musci
e.g. liverworts	e.g. hornworts	e.g. Mosses
like Riccialike	like Anthoceros	like Funaria

- *Class: Hepaticeae:* These are lower forms with an undifferentiated thallus, Rhizoids are unicellular and unbranched, Protonema are absent. Sporophyte is short-lived and simple. However, in some forms it may be differentiated into foot, seta and capsule. Examples are Riccia, Marchantia.
- *Class: Anthocerotae:* Gametophyte is undifferentiated thallus, rhizoids are unicellular and unbranched. Protonema is absent. Sporophyte is differentiated into foot and capsule only. An example is Anthoceros.
- *Class: Musci:* These are higher forms in which the gametophyte is differentiated into stem-like and leaf-like parts showing radical symmetry. Rhizoids are multicellular and branched. Protonema present.

Sporophyte is differentiated into foot, seta and capsule. An example is Funaria.

External Morphology

Since bryophytes lack efficient conducting tissue, they do not become very large. They are small plants forming green velvety patches on moist substratum. They are green due to presence of chloroplasts. The plant body is either thallus or distinguished into stem-like and leaf-like structures. They are attached to the substratum by rhizoids.

Liverworts

The plant body is a thallus which is a gametophyte. It is dorsoventrally flat, showing dichotomous branching. The dorsal surface is marked by a groove throughout, while the ventral surface shows unicellular rhizoids. Reproduction takes place vegetatively, asexually as well as sexually. Vegetative reproduction takes place by production of vegetative reproductive bodies called gemmae. The gemmae eventually separate from the parent plant and grow into gametophytes.

Sexual reproduction takes place by producing antheridia and archegonia which may be on erect branches called antheridiophores and archegoniophores. The antheridia produce biflagellate antherozoids. In presence of water they reach the archegonium, enter the neck and only one fertilizes the egg at the base, to form a diploid zygote developing into asporophyte. The sporophyte remains attached to the gametophyte. Asexual reproduction takes place by production of haploid spores which are produced by the sporophyte. The spore germinates to produce a gametophyte, thus showing alternation of generations.

Mosses

A moss plant begins as a filamentous, green, body called a protonema which produces buds giving rise to the mature moss plants. The moss plant is a gametophyte measuring about 2 to 3 cm in height. It is differentiated into stem-like (cauloid) and leaf-like (phylloid) appendages. There are no true roots.

The plant is fixed to the substratum by multicellular rhizoids. The leaf-like appendages are sessile, spirally arranged on a short stem forming a cluster or rosette at the apex.

Reproduction is mainly sexual, asexual and at times vegetative. At maturity antheridia (male) are borne at the tip of the main branch surrounded by a rosette of leaves, and archegonia (female) are borne at the tip of lateral branch surrounded by a rosette of leaves. The antheridia are club-shaped. At maturity they produce biflagellate, coiled antherozoids.

The archegonia are flask-shaped, each containing one large female gamete called an egg at its broad base. Fertilization takes place in presence of water by the process of chemotaxis, i.e. under the influence of a chemical substance released by the archegonium. The resulting zygote is the beginning of a diploid sporophyte.

The sporophyte is a leafless structure living as a semiparasite on the gametophyte. It consists of a foot which is embedded in the gametophyte and absorbs nourishment, the middle slender, long seta which is used for dispersal of spores, and the apical pear-shaped capsule provided with a lid. It produces haploid spores at maturity. In dry weather, the lid breaks open and the spores are dispersed by wind. Under suitable conditions, a spore germinates to form a protonema. The latter produces buds which give rise to gametophytes, thus completing the life-cycle.

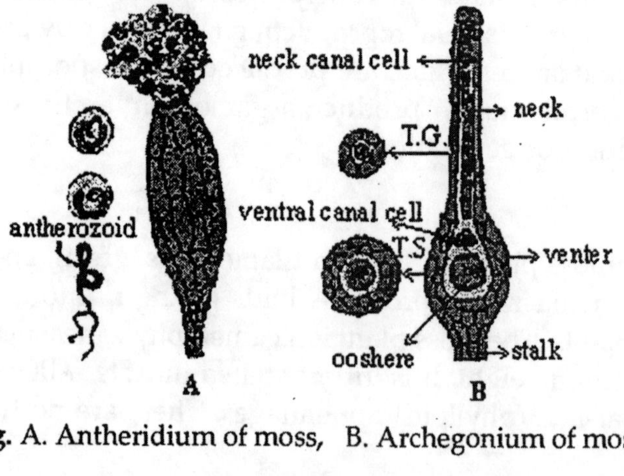

Fig. A. Antheridium of moss, B. Archegonium of moss

Life-cycle Pattern

The life-cycle shows morphologically distinct alternation of generations in which gametophyte is the dominant, independent stage and the sporophyte is a semiparasite or parasite on the gametophyte. The plant is a gametophyte which at maturity produces haploid male and female gametes. The male and female gametes fuse to form a diploid zygote or oospore. The latter develops into a sporophyte. Sporophyte at maturity produces haploid spores which are released in dry weather. Spore germination produces a gametophyte. This successive appearance of two generations alternately one after the other (G-S-G-S) in the life-cycle is called alternation of generations.

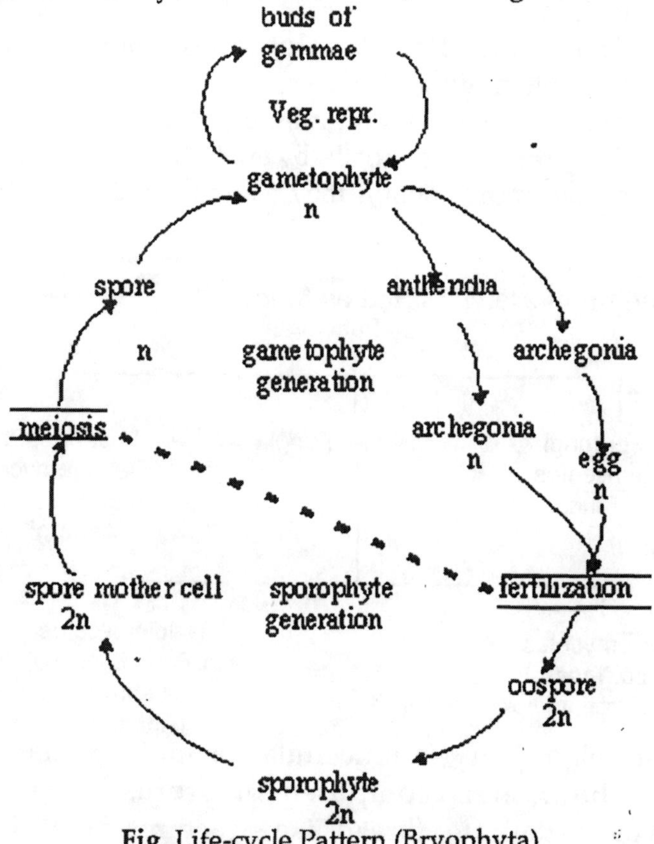

Fig. Life-cycle Pattern (Bryophyta)

Role of bryophytes in nature: (i) Bryophytes provide food for herbivores, birds and other animals. (ii) Bog mosses and peat mosses are used as water absorbing and water retaining material in seed beds and green houses. (iii) They are also used as shock absorbing (packing) material for shipment. (iv) Dead and decaying mosses form humus which increases fertility of soil.

KINGDOM: FUNGI

The fungi are non-green plant-like organisms which are universal in their distribution. They grow in dark and moist habitat and the substratum containing dead organic matter. Mushrooms, molds and yeast are common examples of fungi.

Fungi are distinguished by the following characters:

- They have a definite cell wall made up of fungal cellulose.
- They are without chlorophyll, hence they are heterotrophic.
- They are usually non-motile.
- They reproduce mostly by spore formation. However, sexual reproduction may also take place.

Structure

The fungi are classified as follows:

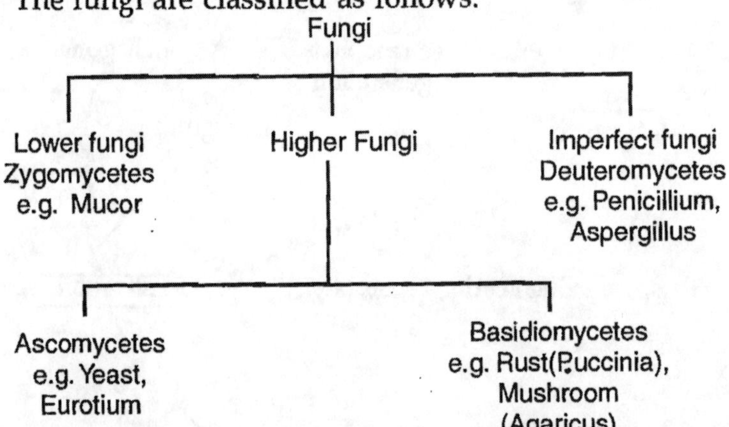

The plant body is unicellular or multicellular. When multicellular, it is composed of profusely branched, interwoven, delicate, thread-like structures called hyphae, collectively called mycelium. Hyphae may be aseptate or septate. When aseptate they are coenocytic, containing many nuclei. Fungi are heterotrophic in nutrition i.e., they may be saprophytic, (living on dead organic matter) or parasitic (subsisting on other living organisms). The parasitic forms may be ectoparasite, living on the body of the host, or endoprasitic, living inside the body of the host. Some forms are symbiotic.

Class: Zygomycetes (lower fungi)

These are filamentous fungi which are terrestrial growing on moist, dead organic matter. They are mainly distinguished by the following characteristics:

- The plant body is called a mycelium; it is made up of interwoven, thread-like structures called hyphae.
- The hyphae are aseptate and coenocytic.
- The sporangia are formed at the apex of aerial hyphae called sporangiophores.
- The spores are numerous and are formed endogenously (i.e. produced within the sporangium).
- Sexual reproduction is isogamous (i.e. the pairing gametes are alike).
- The zygote is unicellular and simple.
- Motile zoospores are produced in some.

Heterothallism

In some fungi there is no morphological sex differentiation, but they show physiological sex differences and said to be having +ve and -ve strains. Such fungi are called heterothallic. (heteros= dissimilar). In these fungi sexual reproduction can occur only between thalli having +ve and -ve strains.

Reproduction

Mucor reproduces asexually as well as sexually. Asexual reproduction takes place by formation of non-motile spores and sexual reproduction takes place by conjugation of similar gametes (isogametes). During asexual reproduction, mycelium gives out upright vertical hyphae called sporangiophores. As the growth proceeds, the tip develops a globular structure called sporangium. At maturity the sporangium breaks open and liberates the spores. On germination, each spore develops into a new mycelium.

Sexual reproduction is isogamousas: it involves the conjugation of two similar gametes. During conjugation the two hyphae regarded as having positive and negative strains come closer (heterothallism). The conjugating hyphae give out club-shaped progametangia which release the terminal part (called gametangia). The gametangia fuse, the middle wall is dissolved and the nuclei fuse in pairs, thus forming a zygote.

The latter develops a thick resisting wall to form a zygospore. Each zygospore, on germination, gives out promyucelium which develops sporangium at its tip. When the sporangium ruptures, the spores are liberated and germinate to produce new mycelia.

CLASS: ASCOMYCETES (HIGHER FUNGI)

These are also called sac fungi distinguished by the following characteristics:

- Mycelium with septate hyphae.
- Asexual reproduction by conidia is a common feature. They are produced at the apex of the hyphae called conidiophores.
- Sexual reproduction occurs by producing sac-like structures called asci, each with usually 8 ascospores, produced endogenously.
- Formation of multicellular complex fruiting body called an ascocarp containing many asci.
- Some ascomycetes, like yeast, reproduce by budding also.

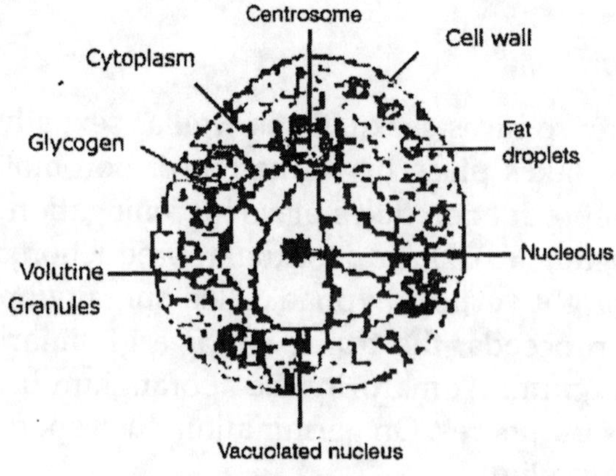

Fig. Yeast

The bodies of ascomycetes may be unicellular as in yeast, or multicellular and filamentous as in powdery mildew. Ascomycetes include yeast, powdery mildew, Claviceps, and most of the blue, green and brown molds.

Reproduction

In most Ascomycetes, asexual reproduction takes place by formation of conidia which are pinched off from the apex of certain specialized hyphae known as conidiophores. e.g. Penicillim. Sexual reproduction takes place by the formation of ascospores within the asci. Some ascomycetes are heterothallic (dissimilar mating types) and other are homothallic (similar mating types). e.g. Claviceps.

Economic Importance

Yeast is used for the production of alcohols in brewery, in production of vitamin B, and for raising bread. Claviceps is used to extract an alkaloid called ergotin which is used in medicine. LSD is also extracted from an ergot fungus. Some ascomycetes like Morchella are edible. Neurospora has been used recently for genetic and biochemical research. Some ascomycetes cause severe plant diseases.

Class: Basidiomycetes (higher fungi)

These are also called club - fungi distinguished by the following characteristics:

- Mycelium with septate hyphae.
- Asexual reproduction is uncommon.
- During sexual reproduction mycelium develops enlarged reproductive, club-shaped structure called a basidium, hence the name.
- Each basidium produces 4 basidiospores at the tip outside (exogenously),unlike ascospores which develop within the ascus.
- In some fungi like mushrooms, a multicellular, complex, fruiting body called a basidiocarp is formed which is often open, sometimes closed. (Basidiocarp = basidia bearing body).

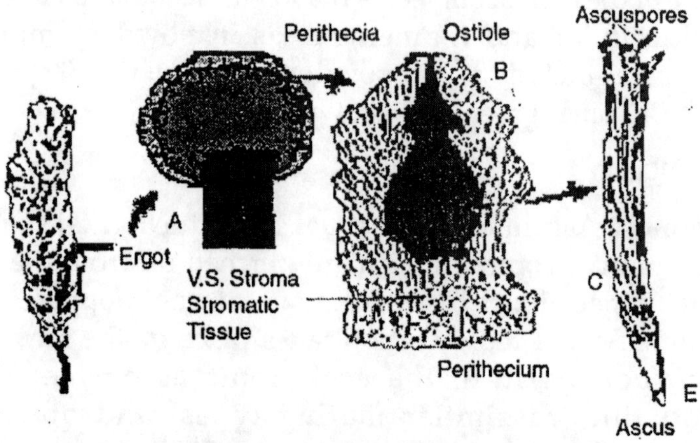

Fig. Reproduction in Ascomycetes (Claviceps)

Basidiomycetes include large and conspicuous fungi like mushrooms, toadstools, bracket fungi and puff balls, as well as parasitic forms like rusts and smuts. The mycelium is made up of profusely branched, septate, multinucleated interwoven hyphae. The major part of the myceluim is underground as in mushrooms, while in parasitic fungi like rusts and smuts it is in the host. Some fungi like mushrooms produce an aerial fruiting body called a basidiocarp made up of stalk and cap bearing gills on the undersurface. Gills produce basidia giving rise to basidiospores, while some parasitic fungi like rusts and smuts produce more than one type of spores in addition to basidiospores outside the host.

Reproduction

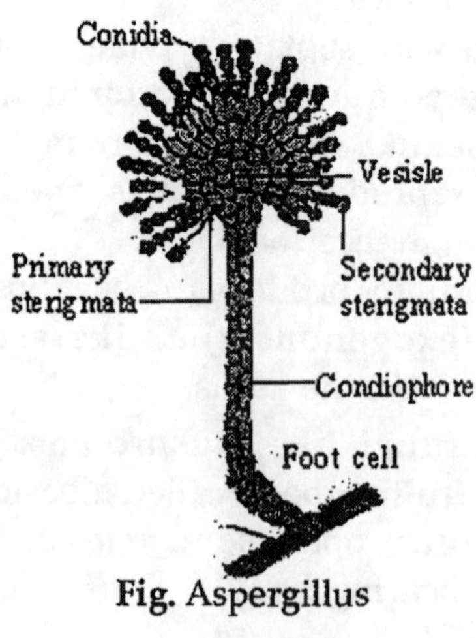

Fig. Aspergillus

Asexual reproduction is uncommon. Sexual reproduction takes place by the formation of basidiospores which are produced by basidia exogenously.

Economic Importance

Some mushrooms are used as food for their delicacy and high nutritive value. Some basidiomycetes are used as a source of medicinal compounds. Some are also used for "recreational" purposes. Some like rusts and smuts cause destruction of crop plants.

DIKARYOTIZATION IN ASCOMYCETES AND BASIDIOMYCETES

Sexual reproduction takes place by the hyphae of two opposite strains growing together in which their cytoplasms intermingles (plasmogamy). Within this fused structure the nuclei of two different strains come together in pairs, but they do not fuse.

This is called dikaryotization. Dikaryotic condition continues for a considerable period of time. New hyphae develop from these fused structures and are called dikaryotic hyphae. These hyphae form a fruiting body called an ascocarp or basidiocarp, characteristic of the class.

Deuteromyucetes

They are also called imperfect fungi because of their imperfect life-cycle, the sexual stage being undescribed. These fungi reproduce asexually,(i.e. by means of conidia) and are closely related to Ascomycetes, while a few are related to Basidiomycetes. Like higher fungi the mycelium is made up of branched, multinucleated, septate, interwoven hyphae. Some of the hyphae penetrate the substratum to absorb nourishment,. while some aerial hyphae are called conidiophores. The latter produce conidia which further give rise to mycelia. Examples of imperfect fungi are Aspergillas and Helminthosporium.

Economic Importance

The fungus Penicillium is used for the production of

antibiotic penicillin. A few species are used to give flavour and texture to cheese. Some are also commercially used for the preparation of organic acid like citric acid, fumaric acid, oxalic acid, etc. Some members cause ring worm and athlete's foot. Some cause diseases in plants. Ascomycetes and basidiomycetes are economically important fungi.

Some of them are used in various industries while other are distinctly destructive. Yeast is used widely by brewers and bakers to prepare alcohol and in raising bread. It is also the main source of vitamin B and hormones like cytokinins. Some mushrooms are used as food for their delicacy and high nutritive value. Fungi like rusts and smuts cause destruction of crop plants while many other cause spoilage of milk, eggs, meat, fish, vegetables, fruit, etc.

DECOMPOSERS AND PARASITIC FORMS

Decomposers

These serve as an important component of the ecosystem. They release hydrolytic enzymes into the substratum and break down the complex organic compound into simpler ones, release minerals and make them available to plants for recycling. *Examples*: Rhizopus, Mucor, Agaricus, Penicillium, If it was not for the actions of decomposers, nutrients would get locked up in the dead matter of organisms. Life on earth without these organisms eventually would become impossible.

Role of decomposers

- They decompose the dead matter of plants and animals and disperse their nutrients, thus acting as scavangers of nature.
- They help in the formation of humus (superficial layer of soil rich in organic compounds) which is important to increase the fertility of soil and healthy growth of plants.
- They play an important role in converting complex organic components into simple forms to continue biogeochemical cycles.

Parasitic forms

These are pathogenic fungi. Unlike saprophytes or decomposers, they do not secrete hydrolytic enzymes, They live on or inside the body of the other living organisms (called hosts) obtain ready-made food from the latter and produce a number of diseases in them. Common plant diseases are potato blight, wheat rust corn smut, and powdery mildew.

Host	Disease	Symptoms	Causativeagents
Wheat Barley Oat	Rust	Rusty spots streaks on stem, lea-sheath andleaves.	Pucciniasp. (basidiom ycetes)
Maize jowar	smut	Black soot-like masses on ear	Ustilagosp. (Basidiom ycetes)
Rye	Powdery mildew	White powdery deposits on leaves	Erysiphesp. (Ascomycetes)
Potato	Late Blight leaves causing willting of leaves and rotting of tubers.	Black spots on under surf aces of the	Phytophthora sp. (phycomycetes)

LICHENS AND MYCORRHIZA

Lichens

It is an obligatory association between an alga and a fungus, which together form a closely integrated unit. The body of a lichen is composed of branching hyphae of a fungus, which harbor algal cells. The fungus gets food synthesized by the alga and the alga in return gets shelter, moisture and minerals absorbed by the fugal partner from the substrate.

Lichens are classified into three principle groups as follows:

- *Crustose lichens:* These lichens form a hard granular crust closely adhering to rocks and tree trunks, e.g. Graphics.
- *Foliose lichens:* These lichens form leaf-like thalli with

lobed margins attached to tree trunks, rocks, walls, etc. They are sometimes referred to as shelf fungus. e.g. Parmelia.

- *Fruticose lichens:* These lichens form much branched shrub-like habit developing fruiting bodies, e.g. Usnea.

Economic Importance of Lichens

Lichens are used in varieties of ways. Some lichens are valuable source of food to wild animals like reindeer. Some lichens are fried and given to cattle as food and to some extent to human beings. Some are used in medicines and other are for preparation of dyes. Litmus is prepared from certain lichens and some are also used in the preparations of cosmetics and perfumes.

Ecological Significance

Lichens growing on rocks disintegrate them to form soil, preparing the ground for mosses and subsequently for higher plants. Thus, they help in the succession of plant communities.

MYCORRHIZA

Symbiotic and usually non-pathogenic associations of fungi with roots of higher plants, are called mycorrhiza. They are observed in 90% of higher plants. The fungus decomposes the organic material in the soil, thus making minerals available to plants, while in return they get sugars, amino acids and other organic substances from the plant. It has been observed that many forest plants die due to malnutrition when the soil lacks appropriate fungi. The best known mycorrhizas are those of forest trees like Pinus and orchids.

CONQUEST OF LAND

Autotrophs are the vital base of food webs present on Earth today. In the ocean, photosynthetic prokaryotes and bacteria provide most of the energy. Today, photosynthetic plants support life on land. Plants evolved from aquatic species not much different from the primitive alge we see today. Plants can be generally defined as multicellular photosynthetic eukaryotes, although there are a few obviously related species which rely instead on alternative modes of nutrition, such as Indian pipe (Monotropa).

Bryophytes were the first organisms to invade land. They developed certain adaptations which allowed them to cope with requirements of a terrestrial environment. They retained some of the characteristics of aquatic species, such as a soft body with the absence of root and vascular systems. The gametophyte is the predominat phase in the life cycle, and the sporophyte is totally dependent on the gametophate. They still need a water environment for the opening of the sex organs and fertilization to occur. Hence, bryophates are sometimes refferred to as the "amphibians of the plant kingdome".

A terrestrial environment entails an entirely different set of environmental conditions compared to an aquatic habitat. Water and nutrients are available only through soil. Oxygen and carbon dioxide needed for respiration and photosynthesis are available in gaseous form in the atmosphere. Aerial structures are exposed directly to atmosphereic conditions like air, light, temperature, wind, e.t.c. Unlike the aquatic environment, the plant does not get natural support or buoyancy from its environment. Thus, in order to survive on land, the terrestrial plants had to develop stratagies, including some of the following characteristics to address these environmental differences:

- An extensive root system for efficient anchorage and absorption.
- Leaves as specialized organs for photosynthesis.
- Specialized vascular tissues such as xylem for conduction of water and minerals from roots to aerial parts, and phloem for the conduction of organic solutes from leaves and other green organs to non-photosynthetic portions of the plant.
- Cuticle over aerial parts of the plant for protection and to prevent surface evapouration.
- Stomata to regulate gaseous exchange with the environment.

The increasing complexity of the plant body was accompanied by "division of lobour" among various organs according to the requirements of the plant body, including metabolic and functional adaptations. Of the twelve phyla of the kingdome plantae, nine of them, known as tracheophytes,.display developed "plumbing" systems for transporting materials throughout the plant. The remaining three phyla contain a very simple or no vascular system and are referred to as non-tracheophates, which replaces the misnomer "nonvascular plants".

Another remarkable change that occurred in the evolution of land plants was the conversion of a predominant haploid, gametophytic generation, as in Bryophyta, to a prodominant diploid, sporophytic generation in the life cycle in pteridophytes, gymnosperms and angiosperms. In bryophytes, the gametophyte is very elobourate, less so in pteridophytes, and greatly reduced in gymnosperms and angiosperms. Fertilization in land plants occurs within the protection of the female sex organ. Evolutionary ancient species produce independant, photosynthetic gametophyte and sporophyte generations.

Spores produced by the sporophytes are of a singular type and develop into a single type of gametophyte bearing both female and male reproductive organs. These are the female archegonium and male antheridium, both found on a single plant. Later, heterosporous plants evolved, with megaspore spores developing into female gametophytes yielding eggs, and microspore spores developing into male gametophytes producing sperm.

PTERIDOPHYTA

Pteridophytes are the simplest of the tracheophytes. They were abundant in the geological past. Today, they are best represented by the ferns, although other non-fern pteridophytes are present in lower abundance. These plants are usually small and herbaceous. They grow well in moist, cool, and shady places where water is readily available.

Distinguishing Characteristics

- The life-cycle shows distinct heteromorphic alternation of generations.
- Diploid sporophyte generation is predominant.

- Sporophyte plant body has true roots, stem and leaves with well developed vascular system.
- Asexual reproduction takes place by spores.
- Most pteridophytes are homosporous; only a few show heterospory.
- Spores are produced in multicellular sporangia after meiosis in spore mother cells.
- Gametophyte is haploid, multicellular, green and an independent structure.
- Sex organs, antheridia and archegonia are multicellular.
- Antherozoids (sperms) are spirally coiled and multi flagellate.
- Opening of sex organs and transfer of male gametes to archegonium for fertilization are dependent on water. Further, fertilization takes place inside the archegonium.

Some common examples of microphyllous pteridophytes (i.e., poorly developed leaves) are club mosses (Lycopodium sp.), and whisk ferns (Psilotum sp.). Examples of megaphyllous (large-leafed) pteridophytes are the ferns Nephrolepis, Osmunda, Pteris, and Salvinia.

CLASSIFICATION OF NEPHROLEPIS

Older classification	New classification	(Pteridophyta)
Kingdom	Plantae	Plantae
Sub-kingdom	Cryptogams	Trachaeophyta
Division	Pteridophyta	Filicophyta
Class	Filicineae	Leptosporangiopsida
Order	Filicales	Filicales
Family	Polypodiaceae	Polypodiaceae
Genus	Nephrolepis	Nephrolepis
Species	exaltata	exaltata

NEPHROLEPIS

Nephrolepis is a common fern which grows in moist, cool

and shaded places in forests, hills, etc. Nephrolepis exaltata is a common species. Various stages in the life cycle of Nephrolepis are completed in two phases. Each phase is known as a generation. The sporophyte generation is the dominant phase while the gametophyte generation is an alternate phase.

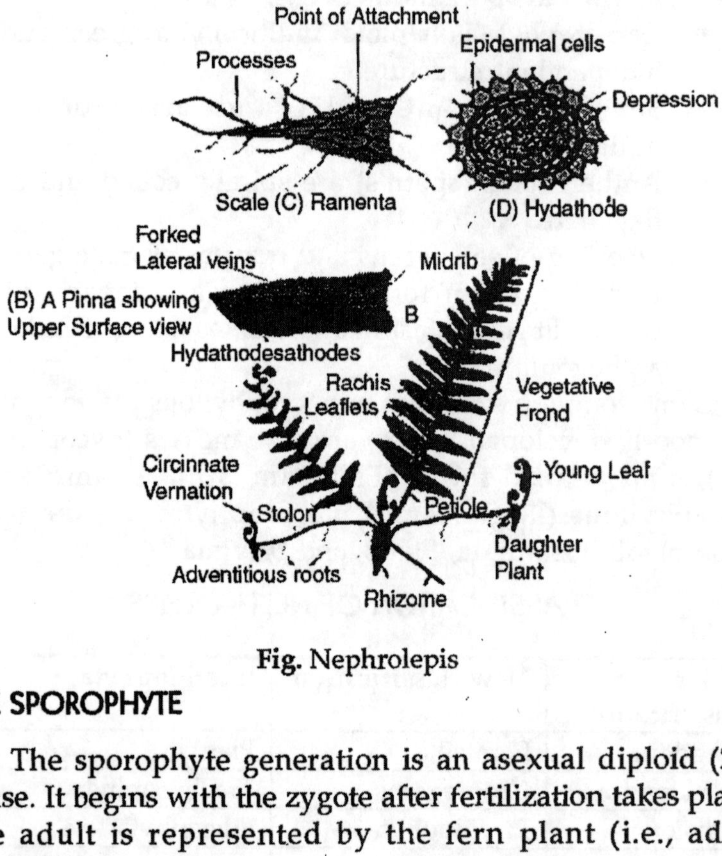

Fig. Nephrolepis

THE SPOROPHYTE

The sporophyte generation is an asexual diploid (2n) phase. It begins with the zygote after fertilization takes place. The adult is represented by the fern plant (i.e., adult sporophyte).

External Morphology

The adult sporophyte (fern plant) is a perennial evergreen herb. It is differentiated into true roots, a stem and leaves (The fungus decomposes the organic material in the soilThe fungus decomposes the organic material in the soilThe fungus decomposes the organic material in the soil). The plant has a well developed vascular system.

- The stem is represented by a rhizome and stolons.
- The main stem is an underground rhizome. It is hard, woody and dark coloured. The rhizome grows obliquely erect and is covered with persistent bases of fallen leaves. It produces green aerial leaves and lateral branches called stolons. The main function of the rhizome is perennation.
- Stolons are the sub-aerial, lateral branches of the rhizome. These spread in all directions and produce new fern plants through vegetative propagation. Each stolon is slender and cylindrical with a smooth surface. Adventitious roots are known to develop from stolons.
- Nephrolepis has well-developed, adventitious, fibrous roots. These are branched and help in fixation and absorption.
- Fern leaves are large, green and aerial. They develop from the rhizome. Young leaves are delicate, coiled like a spring and show circinnate vernation. These are covered with numerous scales called ramenta. Scales protect the young leaves and fall off as the leaves mature.

A fully formed mature leaf is commonly called a fern frond. It is upto 5 feet long and unipinnately compound. The frond has a long petiole. The main rachis bears numerous leaflets (pinnae) arranged in two rows. Each leaflet is linear, broad at the base and tapering towards the apex. It has a midrib with many forked lateral veins. Hydathodes or water stomata are present at the ends of lateral veins on the upper surface of each leaflet. These appear as minute spots and help in removing excess of water and salts from the plant by guttation.

The exuded salts get deposited on hydathodes and form white spots. Hence, hydathodes are called chalk glands. The fern plant is a diploid sporophyte. It shows asexual method of reproduction by spore formation. The spore producing structures are called sori. These develop on the lower surface of leaflets. Leaves with sori are called sporophylls.

Anatomy of a Sporophyll

The transverse section of a fertile pinna passing through a sorus shows the internal structure of the leaflet as well as that of the sorus. The internal structure of the leaflet is differentiated into upper and lower epidermis, spongy mesophyll and vascular bundles. Both the upper and lower epidermis are single layered, non-green and covered with cuticle. Stomata are generally present in the lower epidermis only. Mesophyll tissue is spongy in nature. This is composed of thin walled, green and loosely arranged cells. The vascular bundles of lateral veins are seen in the mesophyll region. Each vascular bundle is concentric and hadrocentric (i.e., it has xylem cells in the centre surrounded by the phloem cells).

The sorus is seen attached to the lower epidermis. It consists of a placenta, numerous sporangia and indusium. The placenta is a fertile tissue of colourless cells. It develops as a cylindrical out-growth from the lower epidermis and bears sporangia and an indusium. Sporangia are the spore producing organs. They develop laterally from the placenta. The indusium is a thin membrane attached at the end of the placenta which protects the sporangia in a young sorus.

The Gametophyte

The gametophyte represents the sexual haploid generation in the fern life cycle. The haploid generation begins after meiosis in diploid spore mother cells. The adult gametophyte is called a fern prothallus. It is produced on moist, cool and shady soil, when a spore (n) germinates under favourable conditions.

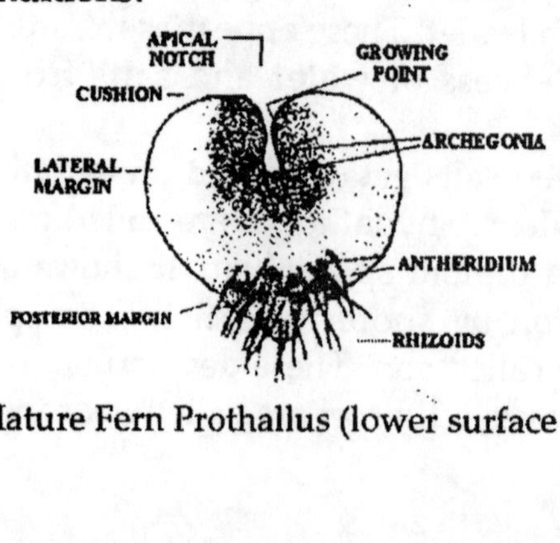

Fig. Mature Fern Prothallus (lower surface view)

The prothallus is very small (about 0.5-2.0 cms in diameter), delicate and almost transparent. It is heart-shaped and flat with dorsoventral symmetry. The prothallus has a deep apical notch. An apical cell (growing point) is situated in the notch. The central part of the prothallus is fleshy and is thick because of a number of cell layers. This is called the cushion region.

The peripheral part has only one cell layer and is called the margin. Numerous hair-like rhizoids develop from the lower surface of the posterior margin region. Rhizoids help in fixation and absorption. The cells of a prothallus contain chloroplasts. Therefore, it is photosynthetic and independent.The prothallus (gametophyte) produces male sex organs (antheridia) and female sex organs (archegonia) during sexual reproduction. These develop on the lower surface facing the soil.

Reproduction

The fern life cycle shows both asexual and sexual reproduction. Asexual reproduction takes place in the sporophyte (diploid) generation while sexual reproduction takes place in the gametophytic generation.

Asexual Reproduction

As stated earlier, this takes place by spore formation. Spores are produced in sporangia which develop in sori.

Sporangium

Each sporangium has a slender multicellular stalk and a body called capsule. The capsule is oval or circular and flattened on two lateral sides like a biconvex lens. The capsule wall is single layered and composed of thin-walled cells. There is a ring of special cells in the wall that runs around the capsule along its median edge. A major part of the ring is formed by the annulus, in which cells have very thick inner walls. The remaining part of the ring is called the stomium region. It is formed by large, thin walled delicate cells. This ring plays an important role in the liberation of spores.

Each sporangium contains 16 diploid spore mother cells. These divide by meiosis and produce 64 haploid spores (n). Nephrolepis is homosporous. All spores are small and kidney-shaped. The diploid sporophyte generation ends with the meiosis.

Liberation of Spores

The spores are liberated from the capsule when the sporangium matures. The sporangium (capsule) wall breaks open in the region of the stomium and the spores are liberated through the opening. This is caused due to movements of the annulus in hot and dry conditions.

Sexual Reproduction

It takes place in the haploid gametophyte generation and involves the formation of specialized male and female sex organs.

Antheridium

This is the male sex organ. Antheridia develop on the lower surface of the prothallus in the region of posterior margin among the rhizoids. Each antheridium is sessile. It has a very small, delicate and globular body. The body wall is called jacket.

It is composed of only three cells: basal cup cell, middle ring cell and apical cap cell. Antherozoids are produced in each antheridium. These are haploid male gametes. Each one has a spirally coiled body with many flagella. The wall of the mature antheridium breaks open in the presence of water, and the antherozoids are liberated on the moist soil.

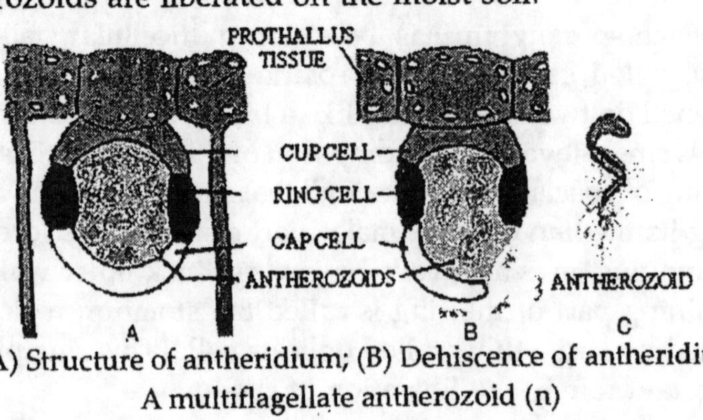

Fig. (A) Structure of antheridium; (B) Dehiscence of antheridium. c. A multiflagellate antherozoid (n)

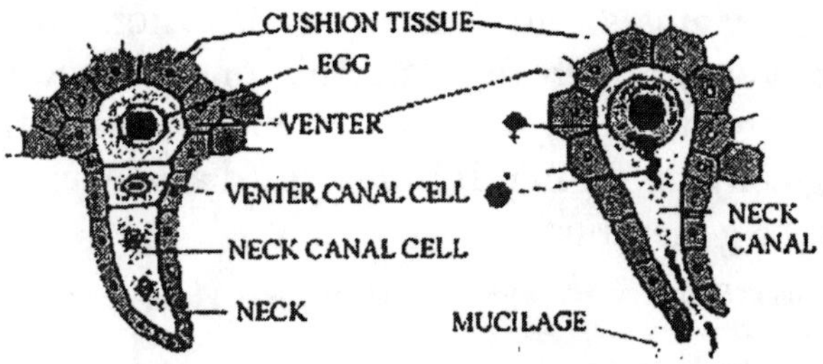

Fig. (A) A Fully Formed Archegonium (B) Fertilization

Archegonium

This is the female sex organ. Archegonia develop towards the lower surface of Prothallus in the cushion region. Each archegonium is sessile with a flask-shaped body. It is differentiated into basal globular venter and a tube shaped neck. The venter is embedded in the tissue of cushion but the neck projects out towards the soil. It is slightly curved. The venter cavity contains a female gamete called an egg (n) cell and a smaller venter canal cell. The neck cavity contains one neck canal cell with two nuclei.

The neck of the mature archegonium opens in presence of water. The venter canal cell and the neck canal cell disintegrate. Their contents form a substance similar to mucilage which comes out through the open neck. Mucilage contains malic acid which attracts the antherozoids for fertilization. This phenomenon is called chemotaxis.

Fertilization

Many antherozoids may enter the open archegonium. However, only one fertilizes the egg. This results in the formation of a diploid zygote (fertilized egg) in the venter cavity. Haploid gametophytic generation ends with fertilization. The diploid zygote (2n) starts a new sporophyte generation and develops into a fern plant (adult sporophyte).

Alternation of Generations

The sporophyte (diploid) and the gametophyte (haploid) generations in fern are produced in alternate succession one after the other. This phenomenon is called alternation of generations. Meiosis and gamete fertilization are the two important events which regulate this phenomenon. As the sporophyte (fern plant) and gametophyte (prothalths) are morphologically dissimilar, it is described as a heteromorphic alternation of generations.

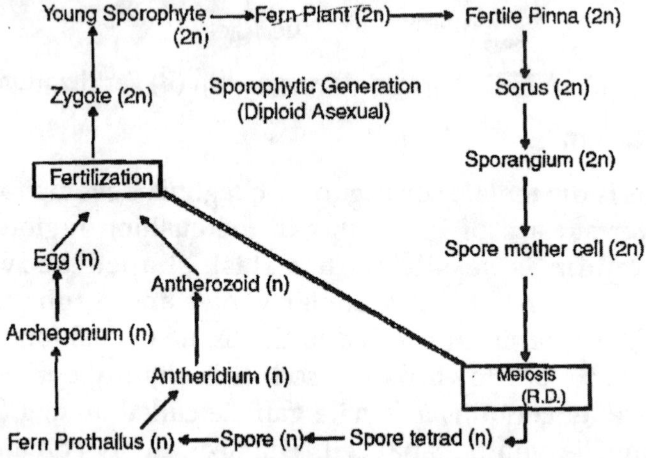

Fig. Schematic Representation of the life Cycle of a fern Showing Alternation of Generations

The sporophyte generation ends when diploid spore mother cells divide by meiosis and produce haploid spores. The spore starts the gametophyte generation. This generation ends when the haploid gametes fuse during fertilization and form a diploid zygote. The zygote again starts a new diploid generation. Alternation of generations is represented schematically in the life cycle of a fern.

GYMNOSPERMS

Gymnosperms (gymnos = naked; spermos = seed) represent a primitive group of seed-bearing plants (spermatophytes) in which the seeds are naked (i.e., without the covering of a fruit wall). This is because in gymnosperms, a closed carpel (ovary) is not formed. The ovules are directly borne on open carpellary leaves (megasporophylls) and hence they are "naked". These develop into naked seeds after fertilization. The term 'gymnosperms' was first used by Theophrastus in 300 B.C. in his book, Enquiry into Plants.

Gymnosperms were most abundant and widely distributed during the mesozoic era. However, they form only a small part of the present day vegetation. There are about 70 genera and 725 species of gymnosperms distributed in tropical and temperate regions. Most of these are the conifers, which are comparatively better-represented today. Others include the cycads and the Ginkgo tree.On the Indian sub-continent, these are found in the form of the coniferous forests in the Himalayas.

Distinguishing Features of Gymnosperms

- The life cycle shows heteromorphic alternation of generations.
- Gymnosperm plants are more advanced than pteridophytes but are more primitive than angiosperms.
- A diploid plant body (sporophyte) is mostly a tree with well developed roots, stem and leaves.
- Two types of reproductive leaves are present: microsporophylls that produce microspores and megasporophylls that produce megaspores.
- The gametophytes are greatly reduced, microscopic and dependent on sporophyte plant body.
- The ovules as well as the seeds are naked.
- The reproductive organs are usually grouped into compact cones or strobili (exception-female reproductive organs in Cycas).
- Pollination is mostly anemophilous (by wind).
- The endosperm (representing the female gametophyte) is a pre-fertilization tissue and is haploid.
- Archegonia are highly reduced and simple.
- Vessels are absent from xylem.

Cycas

living gymnosperms. There are about twenty species of Cycas in the world. Cycas revoluta and Cycas circinalis are two of them.

Classification		
Kingdom	-	Plantae
Sub-kingdom	-	Phanerogams
Division	-	Gymnosperms
Class	-	Cycadopsida
Order	-	Cycadales
Family	-	Cycadaceae
Genus	-	Cycas

External Morphology

Cycas is a slow growing small tree, about 3-6 meters in height. It grows under xerophytic conditions. The plant body is a diploid sporophyte and resembles a small palm tree. It has a tap root system, columnar stem and a dense crown of green leaves. The stem is thick, cylindrical and column-like. It is usually unbranched and has a growing apex (apical bud). The stem surface is covered with a thick, protective armor of the bases of fallen leaves. The tap root (primary root) is very thick and bears secondary roots. Some of the secondary roots are negatively geotropic. These grow upwards to the soil surface and form a special type of root called coralloid roots. These roots form clusters due to repeated irregular branching. Coralloid roots (also called root tubercles) are enlarged, roughly cylindrical and without root-hairs.

They have an algal zone in the region of the middle cortex. Nitrogen-fixing blue-green alga (Anabaena) is present in the intercellular spaces of the algal zone. The alga provides nitrogen compounds to Cycas and in return, it gets space, shelter, water and nutrients from Cycas. Thus, the association is symbiotic.

Cycas has dimorphic leaves, (i.e., two types of leaves): scale leaves and green leaves. Both develop spirally around the stem apex. The spirals of scale leaves alternate with those of green leaves. The scale leaves are small and non-green. The green photosynthetic leaves are between 1-3 meters long and unipinnately compound. They form a crown around the stem apex. Young leaves show circinnate venation (similar to ferns). Each green leaf has a strong petiole and a main rachis with two rows of numerous leaflets (pinnae). Each pinna is long, linear and lanceolate with an entire margin. It has a distinct midrib but lateral veins are absent. Some of the basal leaflets are reduced to spines.

Anatomy of leaflet

A transverse section of a Cycas leaflet (pinna). It has a number of anatomical peculiarities. Some of these are xerophytic characteristics. The general internal organization of tissues is similar to a dorsoventral leaf. The upper epidermis has a thick cuticle but without stomata. The lower epidermis has a thin cuticle and depressed, protected stomata. Sclerenchymatous hypodermis is present under the entire upper epidermis but is restricted to the midrib region of the lower epidermis. Mesophyll tissue is differentiated into upper palisade and lower spongy mesophyll.

A special tissue present in the lamina is called transfusion tissue and is located in between the palisade and spongy mesophyll. It extends laterally from the midrib into the lamina on either side and helps in lateral conduction in the absence of lateral veins. The vascular bundle of the midrib is conjoined and collateral with phloem towards the lower and xylem towards the upper epidermis. It is enclosed in a parenchymatous bundle sheath. The xylem is triangular in shape. It is mesarch with centripetal and centrifugal metaxylem.

Reproduction

ACycas plant is a diploid sporophyte and shows asexual reproduction by formation of spores. Cycas is heterosporous and strictly dioecious (i.e. each plant produces only male or only female reproductive organs). The microsporophylls are produced on the male plant and the megasporophylls are produced on the female plant.

Male Cone

The male cone (i.e. male strobilus) is compact, large (up to 50 cms long) and oval. It is produced singly at the apex of the male plant. Further growth of the stem is sympodial and takes place by the activity of a lateral bud. The male cone consists of a thick central axis on which numerous microsporophylls are arranged in compact spirals. Each microsporophyll is small and flat. It is differentiated into a lower narrow fertile region and a terminal broad, sterile apophysis with a pointed apex. The lower surface (abaxial surface) of the fertile region bears numerous small microsporangia (pollen sacs). They are organized into groups or sori. Each group has 3-6 microsporangia. Each microsporangium contains many diploid microspore mother cells (pollen mother cells). They divide by meiosis and produce haploid microspores (pollen grains). These are liberated from the cone and carried by wind during pollination (wind pollination).

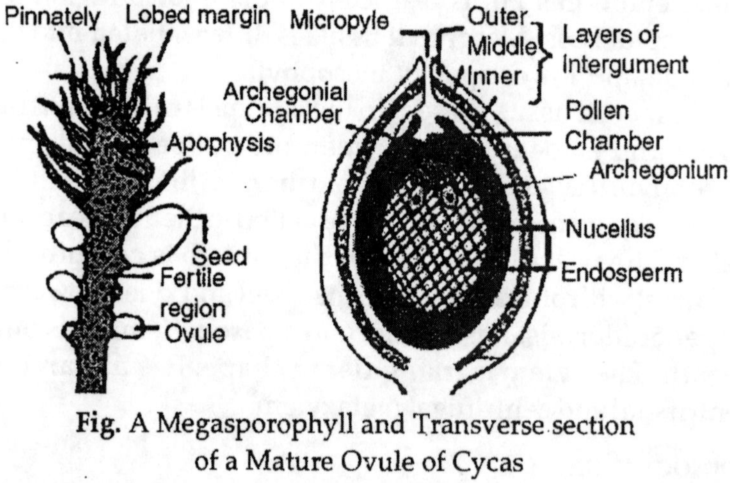

Fig. A Megasporophyll and Transverse section of a Mature Ovule of Cycas

Megasporophylls

These develop in loose spirals around the apex of the stem in female Cycas. Hence, growth of the stem in female plant is monopodial. There is no compact female cone in Cycas. A megasporophyll may be 10-12 inches long. It is narrow in the lower fertile region and is broad and expanded in the upper sterile region called the apophysis. The margin of the apophysis may be serrate or may have long finger -shaped lobes. The fertile region bears 2-10 ovules in two lateral rows. The ovules are sessile, naked and large. The surface of the megasporophyll is covered with soft velvety hair.

Structure of an Ovule (megasporangium)

The ovule is egg-shaped and consists of a nucellus and a single integument. The nucellus forms the main body of the ovule. It has thin walled colourless cells. The integument is thick and differentiated into three layers: the outer and inner layers are fleshy and the middle layer is hard and stony. Micropyle is present in the integument at the apex of the ovule. The nucellus forms a nucellar beak which projects into the micropyle. A cavity develops in the nucellar beak, called a pollen chamber. Pollen grains are deposited here during pollination.

A diploid megaspore mother cell is differentiated in the tissue of a nucellus. It divides by meiosis and forms a linear tetrad of four haploid megaspores. Of these, the upper three megaspores degenerate. The basal (chalazal) megaspore is functional and forms the female gametophyte.

THE GAMETOPHYTES

The two types of spores produce two types of gametophytes.

Microspore and Male Gametopohyte

The microspore produces the male gametophyte. At the time of liberation from the male cone, each microspore contains a 3-celled male gametopohyte: a prothallial cell, a generative cell and a tube cell. Cycas is wind pollinated. The 3-celled pollen grains (microspores) are carried along by wind and are deposited in the pollen chamber of the ovule. The pollen grains germinate here. Each pollen forms a pollen tube which penetrates the nucellus tissue and grows towards the female gametophyte. The pollen tube contains two male gametes (sperms or spermatozoids). These are large, top-shaped and multiflagellate.

Megaspore and female gametophyte

The functional chalazal megaspore produces a multicellular female gametophyte or endosperm in the nucellus. Thus, endosperm is haploid and formed before fertilization in gymnosperms. two to five archegonia develop in the endosperm towards the micropylar end. A cavity called the archegonial chamber is formed in the endosperm near the archegonial necks. Each archegonium contains one haploid egg in the venter cavity.

Fertilization

The pollen tubes pass through the nucellus, reach the archegonial chamber and release the male gametes. A sperm enters the archegonium and fertilizes the egg to form a diploid zygote. If all archegonia are fertilized, 2 to 5 zygotes are formed in the ovule. Formation of a seed: The ovule develops into a seed after fertilization. The integument forms a seed coat. The zygote develops into a dicot embryo differentiated into radicle, hypocotyle, stem tip and two cotyledons. The endosperm is already present in the ovule. It provides nourishment to the developing embryo.

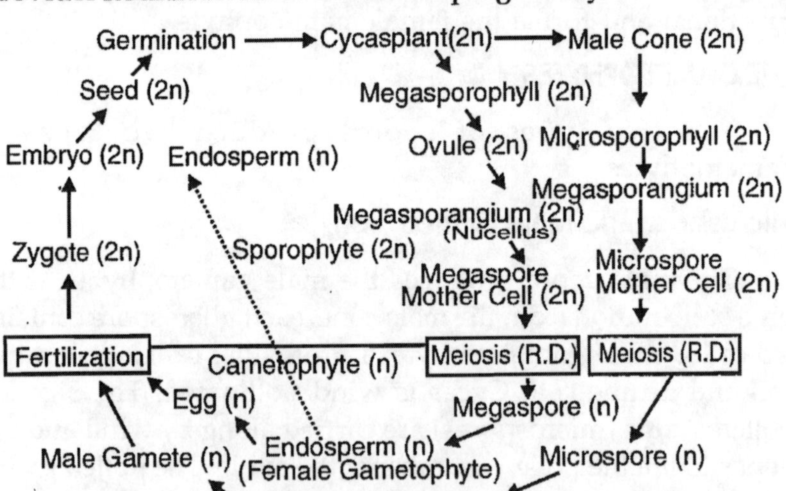

Fig. Schematic Representation of life Cycle of Cycas

In Cycas, 2-5 zygotes are formed in a single ovule. Therefore 2-5 embryos start developing simultaneously. This condition is called polyembryony. However, in a mature seed only one embryo is present. Others degenerate during seed formation. The Cycas the seed is naked, endospermic and dicotyledonous. The seed germinates and produces a new Cycas plant. It may be a male or a female.

Cycas shows a heteromorphic alternation of generations. The dominant generation is diploid, represented by the plant. It ends with meiosis during spore formation. The haploid generation has male and female gametophytes. This ends with fertilization and again a new diploid generation begins.

Angiosperms: Dicotyledons

Angiosperms are the flowering plants in which the seed is enclosed and protected in the pericarp (i.e. the fruit wall). Angiosperms represent the highest evolved and most specialized group in the kingdom Plantae. These are also the most thriving and seen almost all over the world.

Distinguishing Characters of Angiosperms

- The plant body is more complex and better adapted than other plant groups.
- Vascular tissues have specialized conducting elements (e.g., vessels in xylem and sieve tubes and companion cells in phloem).
- The plant body is a diploid sporophyte differentiated into roots, stem and leaves.
- There is the development of a flower as a highly specialized reproductive organ.
- The carpels in the gynoecium have ovary, style and stigma.
- Ovules are enclosed inside an ovary cavity.
- Heterosporous condition.
- Highest reduction in male and female gametophytes.
- Archegonia totally absent in female gametophyte.
- Male gametes are non-motile.
- Pollen grains are deposited on the stigma of the gynoecium during pollination (and not directly in the pollen chamber inside the ovule as in gymnosperms).
- Phenomenon of double fertilization is the characteristic of angiosperms only.

- Endosperm is triploid in nature and develops only after double fertilization is over (post-fertilization tissue).
- Seeds are enclosed inside the pericarp (fruit wall) and thus, the embryo in the seed is very well protected and properly nourished.

Angiosperms are classified as dicotyledonous or monocotyledonous mainly on the basis of following criteria:

	Criteria	Dicotyledonous	Monocotyledonous
(1)	Roots	Tap root system	Adventitious root system
(2)	Venation in leaves	Reticulate	Parallel
(3)	Symmetry of flower	Pentamerous or tetramerous	Trimerous
(4)	Number of Cotyledons in the embryo.	Two	One
Some common examples		Hibiscus (Chinarose), sunflower, mango, pea, bean.	corn, wheat, rice, grasses, onion, Asparagus, Canna, palms, coconut.

Plant Body - External Morphology

The plant body of the Chinarose is a diploid sporophyte. It is an erect, evergreen perennial shrub, about 6-8 feet tall. The plant is differentiated into roots, stem and leaves. The roots form a branched tap root system at the base of the stem. It is an efficient anchoring and absorbing organ.The stem is erect, hard (woody) and cylindrical with a rough surface. Lateral branches develop in acropetal order. Branching of the stem is alternate.

Dicotyledons

Hibiscus rosa-sinensis (China rose)

Classification					
Kingdom	-	Plantae	Series	-	Thalamiflorae
Sub-kingdom	-	Phanerogams	Order	-	Malvales
Division	-	Angiospermae	Family	-	Malvaceae
Class	-	Dicotyledonous	Genus	-	Hibiscus
Sub-class	-	Polypetalae	Species	-	rosa-sinensis

This is one of the most common members of the dicotyledons and shows all the typical characteristics of the group. It is widely cultivated as a garden plant.

Leaves develop on the stem at the nodes in an alternate arrangement. Each leaf is simple and has two free lateral stipules at the base. The leaf lamina is entire, ovate with acute apex and serrate margins. Venation is branched or divergent. Leaves are dorsoventral and deep green. The surface is covered with multicellular epidermal hairs.

Inflorescence

In Hibiscus rosa-sinensis, the inflorescence is a single flower. The inflorescence axis develops in the axil of a leaf and bears a single terminal flower. The joint (articulation) between the axis and the pedicel is distinctly visible.

- The flower is actinomorphic, complete, bisexual, and pentamerous (5-petals). The flower has an epicalyx external to the calyx. An epicalyx is a whorl of 6-7 bracteoles which are small, and green.
- The calyx consists of five green sepals.
- Corolla. Polypetalous (5-petals) are fused with the

staminal tube at the base. The corolla is red, yellow, white or pink.

Reproduction

Hibiscus is heterosporous.

Pollen grains and the male gametophyte: The pollen grains, or microspores, are produced in the anther locules (microsporangia). As stated earlier, anthers are monotocous (i.e., 1-celled). The other locule contains many diploid pollen mother cells (microspore mother cells). These divide by meiosis and produce haploid pollen grains (microspores).

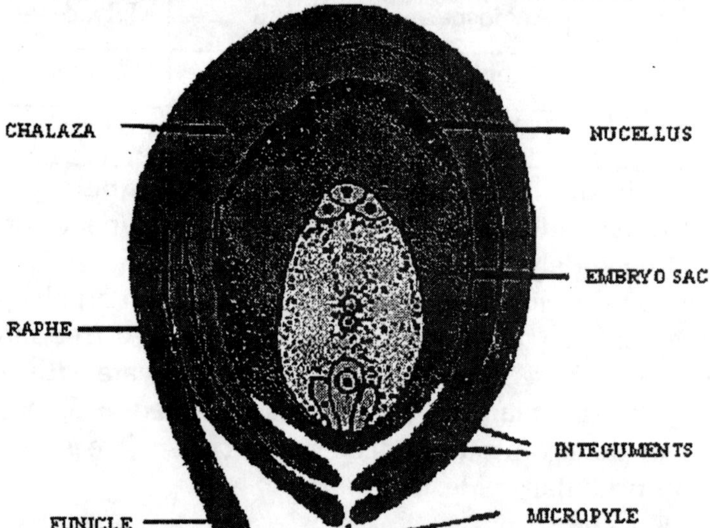

Fig. Cross-section of an Anatropous Ovule Showing Structure of an 8-celled Female Gametophyte

Each pollen is spherical and spiny. The pollen wall has two layers. The inner intine is thin and smooth. The outer exine is thick with a spiny surface and more than one germ pores. Young pollen contains cytoplasm and only one haploid nucleus. The male gametophyte is highly reduced. It is produced within each pollen. A fully formed male gametophyte consists of only three cells i.e., one vegetative cell and two male gametes. The male gametes are non-flagellate and non-motile.

The ovule is stalked. A single diploid megaspore mother cell is differentiated in the tissue of nucellus (megasporangium). This divides by meiosis and produces a linear tetrad of haploid megaspores. Of these, the upper three degenerate. The basal (chalazal) megaspore is functional and produces the embryo sac. It contains 8-celled female gametophyte differentiated into an egg apparatus at the micropyler end, three antipodal cells at the chalazal end and two polar nuclel in the centre of the embryo sac. The egg apparatus consists of one egg cell (haploid female gamete) and two supporting cells called synergids.

Double Fertilization

A number of pollen grains are deposited on the stigmas during pollination. Each one produces a pollen tube on germination. The pollen tube passes through the style, ovary cavity and enters an ovule through the micropyle (porogamy). The tube contains cytoplasm and two male gametes. It penetrates the embryo sac through one of the synergids and releases the two male gametes.

Syngamy

One male gamete enters the egg cell and fuses with it. This is called syngamy or first fertilization, resulting in the formation of a diploid zygote from the fertilized egg. Zygote forms the embryo in the seed. The two polar nuclei fuse together and form a diploid secondary nucleus.

Triple fusion

The second male gamete fuses with the secondary nucleus (2n) to form a triploid primary endosperm nucleus (PEN). This fusion is called triple fusion or second fertilization. The PEN forms endosperm in the developing seed. Syngamy and triple fusion together represent the phenomenon of double fertilization. It is a characteristic of angiosperms only.

The Fruit and Seeds

Soon after the double fertilization, the ovule starts

developing into a seed. The zygote gives rise to a dicot embryo. The nutritive endosperm develops from the PEN. Thus, it is a triploid and post fertilization tissue. The integuments form the seed coat. A mature seed of Hibiscus rosa-sinensis is dicotyledonous, endospermic and hairy. Simultaneously with the development of seeds from fertilized ovules, the ovary also undergoes ripening changes and forms a fruit. The mature fruit is many seeded simple, dry, capsule. The seeds are liberated as the dry capsule dehisces. Each seed produces a new Hibiscus plant on germination.

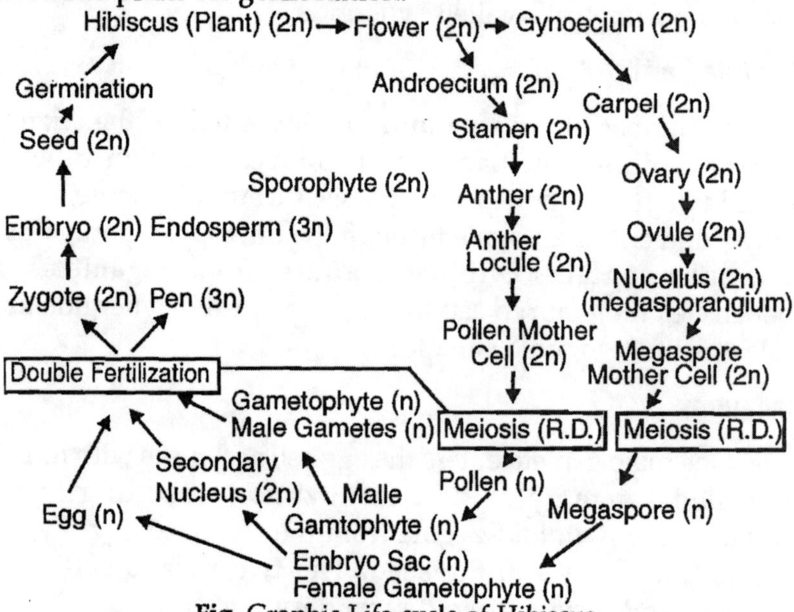

Fig. Graphic Life cycle of Hibiscus

ANGIOSPERMS: MONOCOTYLEDONS

Monocotyledons are angiosperms (flowering plants) with an adventitious root system, parallel venation, trimerous symmetry of flowers and a single cotyledon in the embryo. Sorghum, rice, wheat and corn arethe most commonly cultivatefication.

LIFE CYCLE

The life cycle of sorghum is predominantly sporophytic and shows heteromorphic alternation of generations.

Kingdom	-	Plantae
Sub-kingdom	-	Phanerogams
Division	-	Angiospermae
Class	-	Monocotyledones
Series	-	Glumaceae
Family	-	Gramineae
Genus	-	Sorghum
Species	-	Vulgare

Plant body- External Morphology

The sorghum plant is a diploid sporophyte. It is an herbaceous, tall annular plant which is sown during the monsoons. The plant is differentiated into roots, stem and leaves.

- The root system consists of two types of roots; primary and adventitious. The primary roots, called seminal roots develop first and penetrate vertically downwards in the soil. The adventitious roots develop from basal nodes. These are fibrous, branched, well developed and also function as stilt roots to support the tall plant.
- The stem is erect, solid, up to 10 feet tall and with distinct nodes and internodes. Few basal internodes are short. The internodes are smooth and cylindrical with a longitudinal furrow along one side.
- The leaves are simple, sessile and alternate. The leaf blade (lamina) is long and lanceolate with an acute apex. The surface is rough and hairy. Venation is multicostate and parallel, but with a very well developed mid-rib. Leaf base is long and sheathing. There is a short membranous ligule at the junction of the sheathing base and the lamina.

Inflorescence

In sorghum, inflorescence is a terminal panicle (cluster).

and develops as a more or less compact head. It consists of a profusely branched inflorescence axis (rachis). The ultimate branchlets of the rachis bear one to many pairs of spikelets. In each pair, one spikelet is sessile and perfect (bisexual) while the other is pedicelled and staminat.

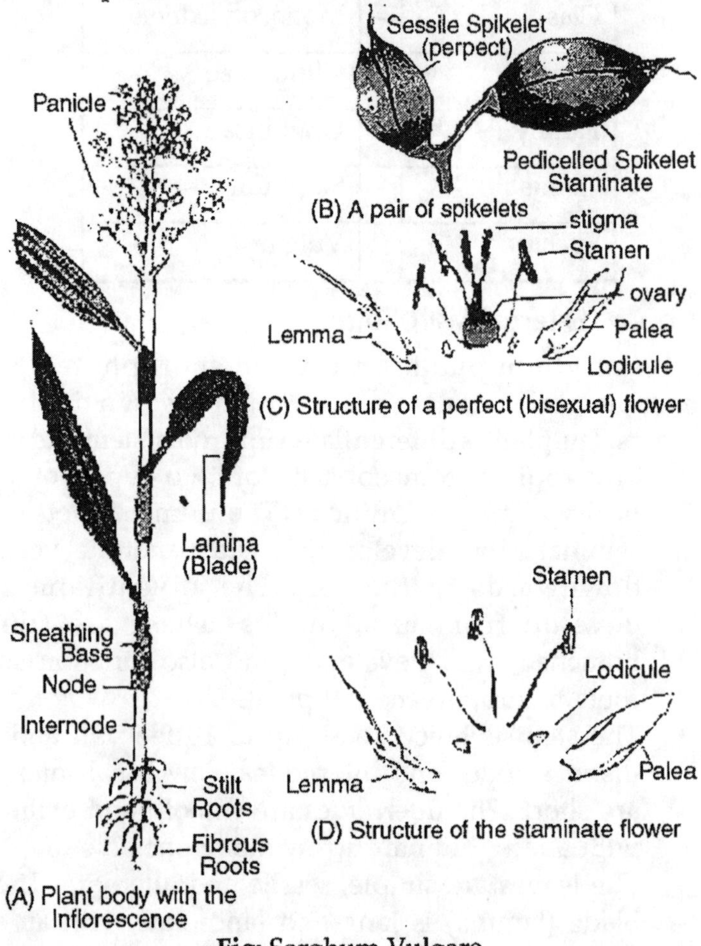

Fig: Sorghum Vulgare

Perfect (bisexual) Spikelet

It is sessile and comparatively broader and larger. The axis of the spikelet is short and bears two flowers (florets). The lower flower is sterile and non-functional (i.e., without sex organs). The upper flower is perfect (i.e. functional and bisexual). Each flower has two bracts; one called the lemma and the other, the palea.

Structure of a perfect flower

Perianth is represented by two colourless, small, broad and hairy perianth members called lodicules. Androecium has three free stamens. Anthers are bilobed, bithecous and versatile. Gynoecium is monocarpellary. Ovary is superior, with one ovule in a basal placentation. There two long and free styles. Stigmas are short and bushy. The pedicelled staminate spikelet also has florets. The lower one is sterile while the upper one is staminate. It has three stamens and two lodicules, but a gynoecium is absent. Each floret has a lemma and a palea. The spikelet is enclosed in two glumes-outer and inner.

Reproduction: Sorghum is Heterosporous

The micropores or pollen grains are produced in the anther locules (microsporangia). Each anther locule produces many diploid pollen mother cells. These divide by meiosis and produce tetrads of haploid pollen grains. Young pollen grains are uninucleate. The male gametophyte is highly reduced in size and is produced within the microspore (pollen) wall. A fully formed male gametophyte consists of only three cells namely a vegetative cell and two male gametes. The mature pollen contains a 3-celled gametophyte at the time of pollination. The pollen wall has two layers, intine and exine. Exine is thick, smooth and has one germ pore.

- The ovule (megasporangium) contains a nucellus is surrounded by two integuments. One diploid megaspore mother cell is differentiated in the nucellus of a young developing ovule. This divides by meiosis and produces a linear tetrad of haploid megaspores. The upper three megaspores degenerate. The basal one is functional and produces the embryo sac in the nucellus. The embryo sac contains an 8-celled female gametophyte differentiated into an egg apparatus, polars and antipodals..
- The egg apparatus consists of one egg cell and two synergids at the micropyler end in the embryo sac. The antipodals are three cells at the chalazal end. Two polar nuclei are present in the centre of the

embryo sac. They fuse together and form a diploid secondary nucleus.

- Pollination is anemophilous, i.e. by wind.

Double Fertilization: It is similar to that described for Hibiscus As a result of double fertilization, the ovule develops into a seed and the ovary forms the fruit.

- The fruit is a one-seeded caryopsis. It is oval or spherical, white and commonly called grain.
- The seed is monocotyledonous and endospermic. The seed coat is fused with the pericarp. The embryo in the mature seed has a single cotyledon (scutellum), radicle enclosed in coleorhiza and plumule covered in coleoptile.

The grain on germination produces a sorghum plant.

Alternation of generations

The life cycle of sorghum shows heteromorphic alternation of generations. The sporophytic generation is predominant. The male and female gametophytes are highly reduced. Meiosis and fertilization are the two important events which regulate the alternation between diploid and haploid generations. This is evident from the graphic representation of the stages in the life cycle of sorghum.

VASCULARIZATION

Evolution of vascular tissues in plants was one of the essential adaptations to terrestrial habitats. Vascular tissues are the specialized conducting tissues, xylem and phloem. Plants with vascular tissues are grouped as the Tracheophyta and include Pteridophyta, Gymnosperms and Angiosperms.

Xylem

It is a specialized tissue for conduction of water and minerals from roots to all other parts of the plant. Xylem is a complex and heterogeneous tissue composed of four types of elements namely (a) tracheids,(b) vessels,(c) fibers,and (d) xylem parenchyma. The first three elements are dead and lignified. Hence, xylem is described as a dead tissue.

Phloem

It is a living, complex and heterogenous food-conducting tissue. It is composed of:

- Sieve tubes,
- Companion cells,
- Phloem fibers
- Parenchyma.

In plants, vascular tissues are organized into strands called vascular bundles. These are spread over all parts of the plant and form the vascular tissue system. In gymnosperms and dicotyledons, cambium is also present in the vascular bundles.

Steles

The vascular tissues are organized into steles. The vascular bundles, with or without other non- vascular tissues, when surrounded by the pericycle and the endodermis are referred to as steles.

Main Types of Steles and their Evolution

- *Protostele:* This is the simplest and most primitive type of stele in which xylem forms the central cylinder and is completely surrounded by phloem cells. Pith is totally absent. Common in pteridophytes like Psilotum, Lycopodium, etc.
- *Siphonostele:* It has evolved from the protostele. In siphonostele, pith is present in the centre of the xylem cylinder. The phloem is external to xylem.
- *Solenostele:* A Cbroken at one point at a time due to the non-overlapping leaf gaps is called solenostele
- *Dictyostele:* Siphonostele in which many overlapping leaf gaps occur at the same level is called a dictyostele, or dissected siphonostele. In a transverse section it appears broken into many smaller or larger arcs. Each arc is known as a merestele (e.g. fern rhizome). Each meristele is completely and independently surrounded by the pericycle and the endodermis. This is the most highly- evolved stele type in Pteridophyta.

- *Eustele:* The characteristic stele present in dicot and gymnosperm stems is called a eustele (i.e." true" stele). It is also a dissected siphonostele in which the vascular cylinder is dissected into separate vascular bundles arranged in a ring. The entire ring of vascular bundles is surrounded by the pericycle and the endodermis.
- *Atactostele:* This is a complex type of stele in which numerous vascular bundles remain irregularly scattered in the ground tissue (atactos means without any order). Atactostele is a characteristic of monocot stems. Here, due to the scattered vascular bundles, the cortex and stele are not demarcated. Similarly, the endodermis and the pericycle are not differentiated.

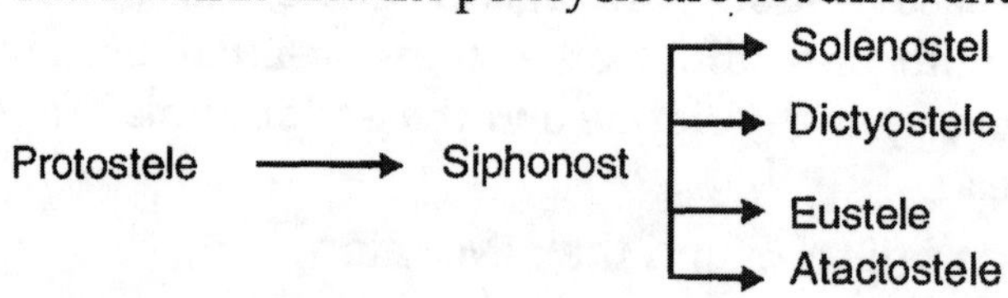

Development of Seed Habit

Evolution of seed structure was one of the most essential requirements of the terrestrial habit. This has equipped the Phanerogams (spermatophytes) that is, seed bearing plants) with the best and most efficient means of propagation.

Distinguishing Characteristics

The basic characteristics required in plants for development of seed structures may be summarized as follows:

- Heterospory
- Only one megaspore per megasporangium
- Extreme reduction in both male and female gametophytes.
- Retention of the female gametophyte within the megasporangium on the parent plant for proper parental care of new embryo.
- connection between female gametophyte and the wall of the megasporanguim.

- Protective integument around megasporangium
- Siphonogamy
- Temporary suspension of growth of embryo (i.e. dormancy period).

Phanerogams possess all these characteristics and hence can produce seeds.

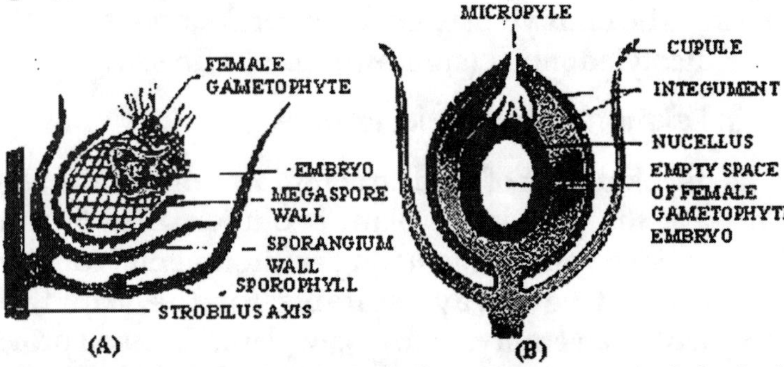

Fig. Stages in the Development of seed Structure A. Selaginella. Female Gametophyte with the Embryo is within Megasporangium. B. Longitudinal Section of the ovule of a fossil seed fern e.g. Laginostoma

Pteridophytes are non-seeded plants. Most of them are homosporous. However, a few pteridophytes are heterosporous e.g. Selaginella, Isoetes, Marsilea, Azolla, Salvinia,. Some of these possess some of the characteristics needed for seed formation. Gymnosperms and angiosperms are the seed bearing plants (Phanerogams). In these plants, megasporangium (nucellus) is surrounded by the protective integument and is called an ovule. In gymnosperms, it is directly borne by the megasporophyll (open carpel) and is called a naked ovule. It develops into a naked seed after fertilization.

The carpel in angiosperms is differentiated into ovary, style and stigma. Ovules are encased in an ovary cavity (closed carpel). After double fertilization, the ovule forms a seed and the ovary forms the fruit. Hence, seeds are very well protected and the embryo in the seed gets proper parental care, protection and nourishment. During seed development, the zygote forms an embryo, and integuments form a seed coat. In gymnosperms, the female gametophyte acts as endosperm. Hence, it is a haploid and pre-fertilization tissue. In angiosperms, endosperm is a triploid and post-fertilization tissue.

A mature seed of gymnosperms is endospermous and the embryo may have two cotyledons (e.g. Cycas) or many cotyledons (e.g. Pinus). In angiosperms, mature seed may be endospermic (e.g. sorghum, corn) or non-endospermic (e.g. pea, bean). The embryo may be monocotyledonous (sorghum, corn) or dicotyledonous (sunflower, pea, Hibiscus).

Seeds in Relation to Terrestrial Habitat

The development of seed structure was one of the biggest advances in the process of plant evolution. The terrestrial habitat provides widely fluctuating environmental conditions, which may not be always suitable for the growth and development of an embryo into a new plant. The seed provides necessary protection and care to the embryo. In seeds, growth of an embryo is takes place only when external conditions are favourable. This increases the chances of survival of the new individual.

Seeds of different plants are adapted for dispersal by wind, water, animals, etc. Hence they act as highly efficient vehicles of dispersal and propagation. The resistant seed coat protects the inner embryo from extremes of external conditions e.g. fluctuations in temperature, water, oxygen, light, etc. Moreover, the internal structures of a seed, especially the embryo are physiologically adapted to sustain prolonged adverse conditions of water, temperature, etc.

DEVELOPMENT OF FLOWER AND FRUIT

Development of Flower

The flower arises as a flower primordium in an axillary or terminal position. Soon, the primordium differentiates into a pedicel (flower stalk) and the floral axis (receptacle). The floral leaves arise as the lateral appendages from the axis at the successive nodes in acropetal order. These appendages arise as small protuberances called primordia and grow into various floral organs.

More commonly the sequence of development, from the outside towards the centre, is sepals, petals, stamens and carpels. These primordia increase in size by terminal and marginal growth in a manner similar to that of vegetative leaves. The development of most parts of the flower is complete by the time the flower opens. However, some parts continue to grow even after that. For example, the carpels (gynoecium) continue to grow as the ovary develops into a fruit.

Development of fruit

Fruit is defined as the ripened ovary. Normally, a fruit develops from the ovary after fertilization (true fruit). The ovary wall undergoes the ripening changes and forms the fruit wall, called the pericarp. The ovules develop into seeds. In many fruits, along with the ovary, certain other parts of flower or inflorescence (e.g. thalamus, calyx, inflorescence axis, etc) also participate in fruit formation (false fruit).

During ripening (maturation), various structural, physiological and biochemical changes take place in the fruit wall (ovary wall). Fertilization provides the stimulus necessary for initiating these changes. Some plant hormones are probably released in the ovary by the pollen tube during fertilization. Also, the developing seeds presumably release hormones enhancing the ripening changes. Auxins, gibberellins and kinins are believed to play an important role in the process.

The mature fruit may have one or more seeds. The pericarp at maturity may remain soft and fleshy or may become dry. It may be dehiscent or indehiscent. Since flowers with closed ovaries are present only in angiosperms, fruit is formed only in angiosperms (flowering plants).

Chapter 10

X-ray Analyses of Proteins and Nucleic Acids

PROTOPLASM

Living matter is often referred to as protoplasm. This is a loose definition; it would be more exact to speak of the living cell and its contents, but the word protoplasm is shorter and easier. Protoplasm contains many types of structures and a wide variety of types of molecules. Two types or families of molecules, both found in protoplasm, are characterized in terms of their physical shape and spatial arrangement.

One of these families, the proteins, occurs abundantly in all living matter and forms an important part of all animal diets. The other, the nucleic acids, is found only in relatively small quantities but is very important for all life. Both families of compounds are high polymers, that is, molecules composed of many small units called monomers. Proteins and nucleic acids are large molecules with molecular weights as high as 10^7. These two families of molecules differ chemically in that they are polymerized from different types of monomers.

Some idea of the protoplasmic roles of proteins and nucleic acids may be obtained by considering a typical cell. The structures shown are based on the results obtained with a variety of techniques. There are two major subdivisions: the nucleus and its contents, often called nucleoplasm; and the remainder of the cell, sometimes called cytoplasm. Both the nucleus and the cytoplasm are surrounded by membranes, as are also the smaller formed elements such as the mitochondria and the Golgi bodies. These membranes consist partly of proteins and partly of other compounds called lipids (for example, fatlike compounds). The membranes not only give physical form to the structures they surround, but also can use energy to transport molecules actively against electrochemical gradients.

Proteins are found in abundance within the formed elements, the contractile elements, the nucleus, and the endothelial reticulum. Some of these proteins serve structural purposes and others act as catalysts for the numerous biological reactions which must occur at controlled rates if the cell is to live. Other proteins are found in the liquid parts of the cytoplasm and nucleoplasm; these are probably mostly enzymes, although some may be concerned with the osmotic balance of the cell. The number and variety of different proteins seem almost limitless.

The nucleic acids are also found both in the cytoplasm and in the nucleoplasm. Those in the cytoplasm are all of a type called RiboNucleic Acids (RNA); they probably exist primarily as small collections of RNA along the endothelial reticulum. Within the nucleus, there are two types of nucleic acids. One type is RNA, similar to that found in the cytoplasm; the other type, the Deoxyribose Nucleic Acids (DNA), occurs only in the nucleoplasm. Evidence was presented to show that DNA is associated with the transmission of genetic information. During cell division, the DNA in the nucleus of most plant and animal cells is organized into long threads, called chromosomes. The chromosomes contain both protein and DNA; they are referred to as nucleoproteins.

The RNA, in contrast with the DNA, is not restricted to the chromosomes or the cell nucleus. It is believed that DNA controls the synthesis of RNA which in turn controls the synthesis of proteins. Thus, both DNA and RNA act as biological catalysts, controlling ultimately the synthesis of protein-enzymes. These, in turn, control the rates of most other chemical reactions within the cell.

There are many other types of molecules within the cell besides proteins and nucleic acids. In the typical living cell, there are more molecules of water than of any other compound. Water makes up as much as 80 per cent of the cell weight in some cases. The fatlike molecules, that is, lipids, in connection with membranes but are by no means restricted to the membranes. Rather, lipids are found in varying roles throughout the cell. Those in fat globules are used to store energy, and a few lipids are hormones. However, the role of most lipids is unknown. The typical lipid has a molecular weight between lipids. Lipids are one general class of molecules found in all cells. The phospholipids are a part of 100 and 1,000, which is small compared to those of proteins.

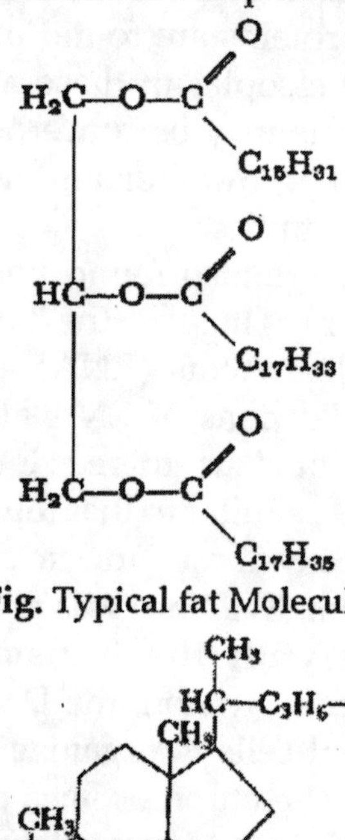

Fig. Typical fat Molecule

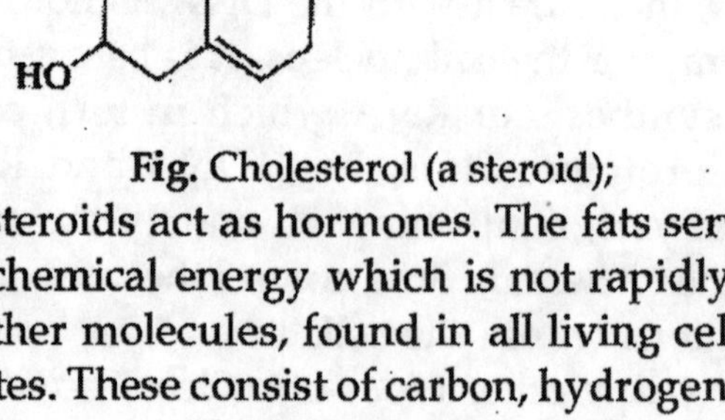

Fig. Cholesterol (a steroid);

Many steroids act as hormones. The fats serve as storage depots for chemical energy which is not rapidly available to the cells. Other molecules, found in all living cells, are called carbohydrates. These consist of carbon, hydrogen, and oxygen atoms, the latter two always occurring in the same ratio as in water. The carbohydrates sometimes are found as small ring or chain structures called monoses, or simple sugars having three to seven carbon atoms. The most common number is six carbons, as exemplified by the common sugars glucose and fructose. Carbohydrates are also found as compound sugars (or dimers) such as sucrose, and as long polymerized chains of monoses such as starches and celluloses.

Of the various compounds found in the cell, the proteins and nucleic acids have been singled out and in the three following ones for a number of reasons. First, their structure and their action have been studied by means of the methods of physics and physical chemistry. Second, their importance both for cell life and reproduction is understood. In addition, they present a complexity and diversity which challenges man's abilities to investigate them, as well as a simplicity and uniformity in over-all plan which cannot but fill one with awe.

Proteins

To recapitulate briefly, proteins are one of the major classes of compounds found in all living matter. Enzymes are protein catalysts controlling the rates of many biological reactions. Muscular contraction depends on proteins, and active transport across cell membranes appears to be a lipoprotein function. Many physical means have been used to study protein structure; currently accepted ideas lean heavily on the results obtained from studies of X-ray diffraction. Before examining the X-ray data, the chemical composition of proteins will be briefly considered.

Proteins are natural high polymers built from small blocks, called amino acids. These are molecules with an average molecular weight of about 120, each of which has an organic acid group, and a basic amino group, $-NH_2$.

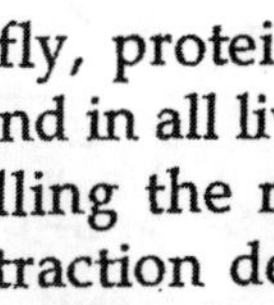

The acid and basic groups are attached to the same carbon leading to the form, where R is either H or any of a number of different organic radicals. This form is called an á-amino acid.

Amino Acid	Structure
1. Glycine	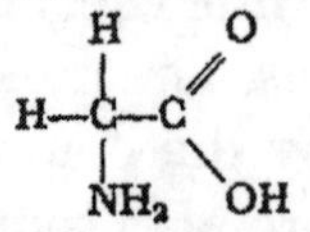
2. Alanine	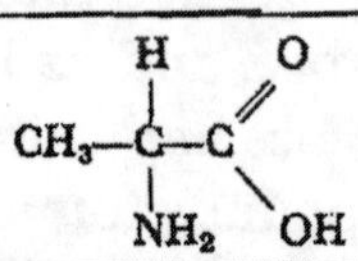
3. Valine	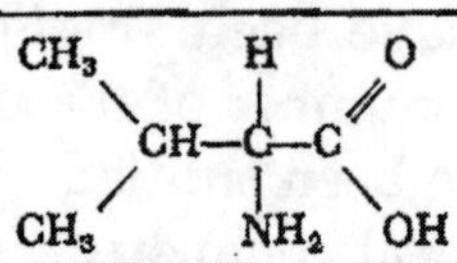
4. Leucine	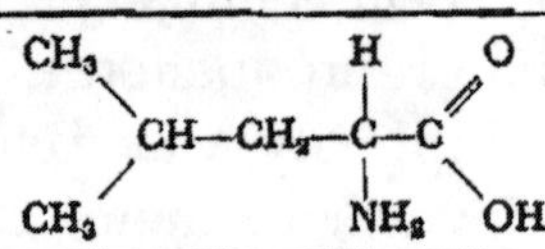
5. Isoleucine	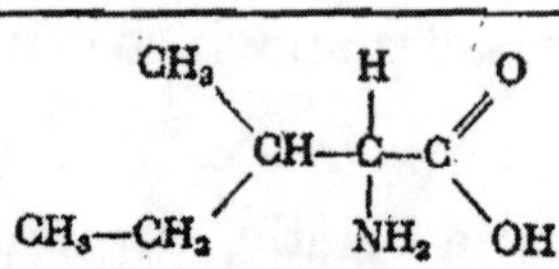
6. Serine	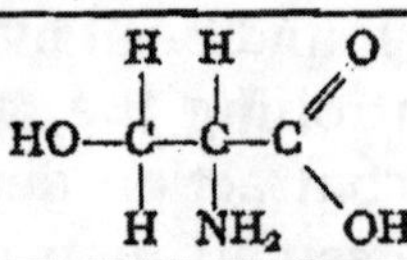
7. Threonine	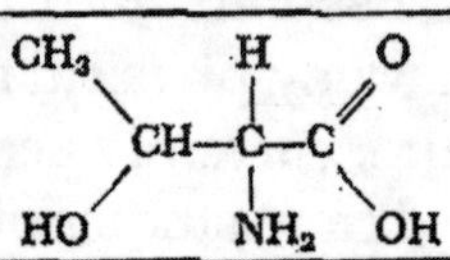
8. Phenylalanine	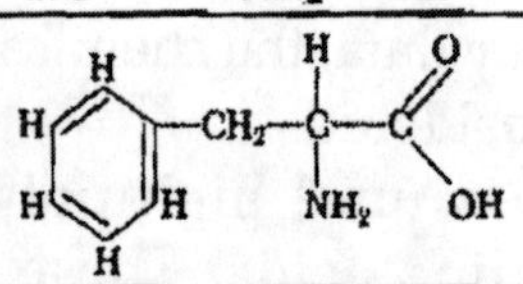
9. Tyrosine	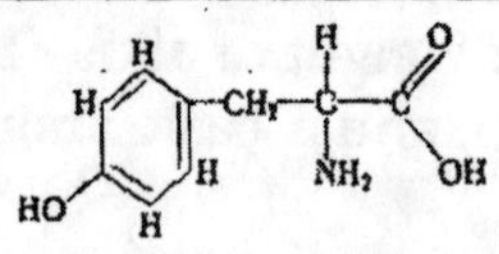
10. Trytophan	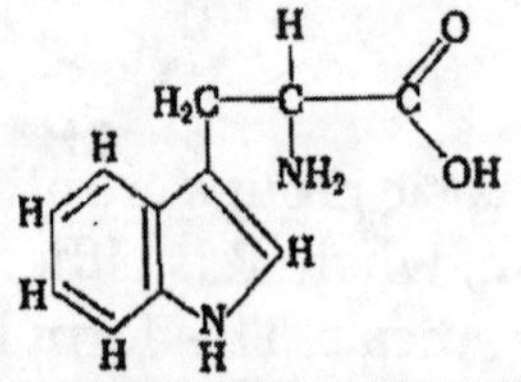

Amino Acid	Structure
11. Cysteine	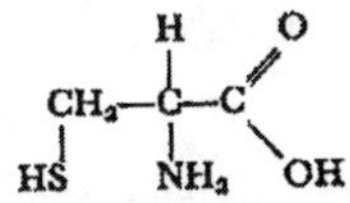
11a. Cystine	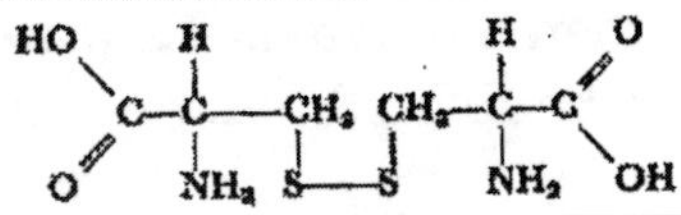
12. Methionine	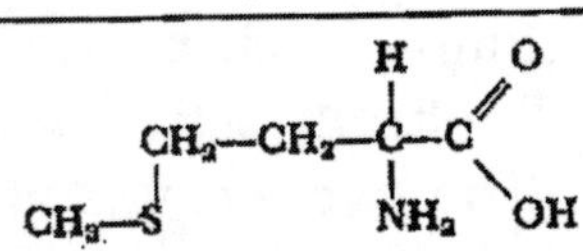
13. Aspartic Acid	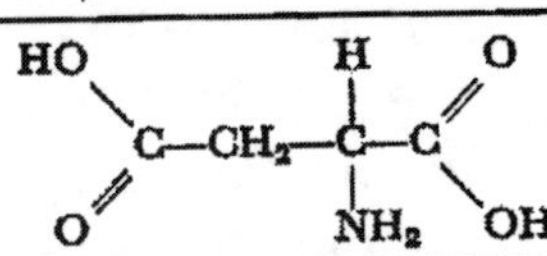
14. Glutamic Acid	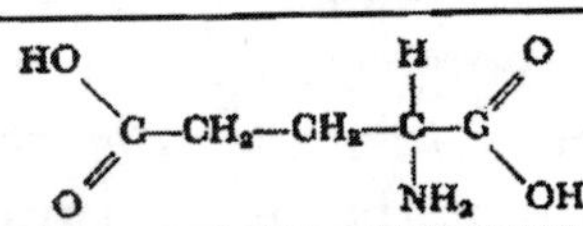
15. Lysine	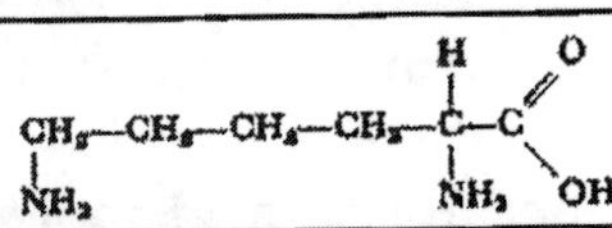
16. Arginine	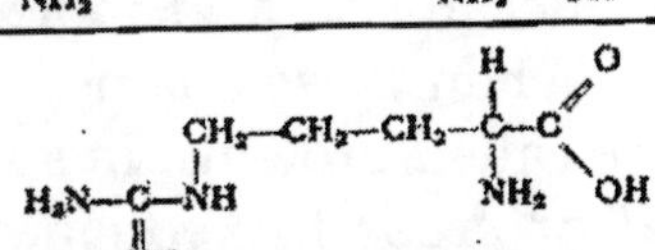
17. Citrulline	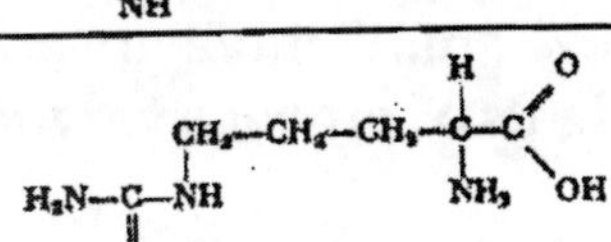
18. Histidine	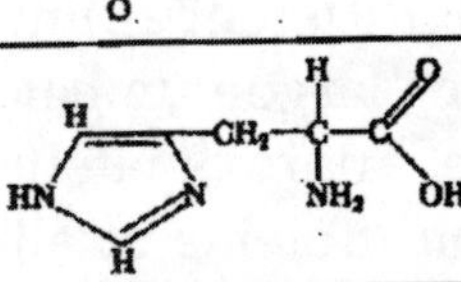
19. Proline	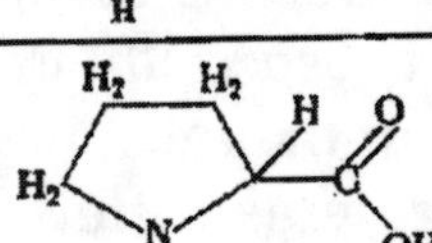
20. Hydroxyproline	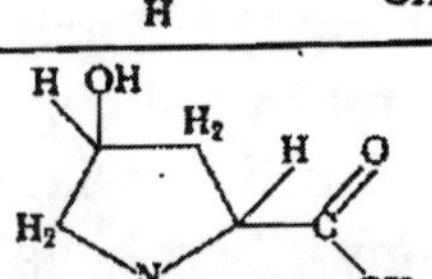

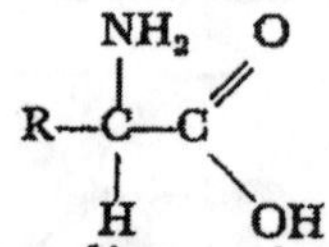

If one makes a three-dimensional model of such an á-amino acid, there are two steric arrangements of the á carbon which cannot be rotated into one another (except when R is a proton).

A carbon atom, which is sterically asymmetric in this fashion, is called optically active because in solution it rotates polarized light. The two stereoisomers are labeled D and L for dextro-rotary and levulo-rotary, respectively. Most test tube syntheses give equal amounts of the D and L isomers, but most living cells produce only one variety or the other. With very few exceptions, the amino acids polymerized into proteins in the living cell are all L-á-amino acids.

Two amino acid molecules may react to eliminate a molecule of water, thereby forming a peptide bond. The peptide bond so formed is very stable and the molecule is called a dimer or dipeptide. It is then possible to attach this molecule to other amino acid molecules, forming chains, or polypeptides. When these chains include 50 or more amino acids, they are called proteins. In some cases, the chains may be branched or cross linked; many proteins contain, in addition, a few small molecules other than amino acids. Molecular weights of proteins vary from several thousand into the millions.

The number of different amino acids conceivable has no known limit. A very large number have been synthesized in test tubes. Of these, approximately 20 L-á-amino acids make up almost all of the proteins of all living cells. Other amino acids occur in nature, especially in bacteria, but they are the exceptions rather than the rule.

The various proteins differ from one another in the number and order of the various amino acids in the polypeptide chains and in the configuration of these chains. Although the detailed order is known for only a few small proteins, pieces of many others have been studied to determine the order of the amino acids.

NUCLEIC ACIDS

As recently as 1950, many scientists regarded proteins as the fundamental "stuff" of life, controlling reactions, contracting, transmitting genetic information, and reproducing themselves. This view has given way to one which assigns proteins to a more restricted role but emphasizes the importance of the nucleic acids in protein synthesis and genetic transmission of information.

The nucleic acids are high polymers, just as are proteins, but are given a separate name because the monomers from which nucleic acids are built are not amino acids. Consisting of different structural units, the nucleic acids differ significantly from proteins in their physicochemical properties, as well as in their biological action.

The monomers from which nucleic acids are polymerized are called nucleotides. Each nucleotide consists of a nucleoside condensed by the elimination of a water molecule, with a phosphate group. A nucleoside in turn is the condensation product of a five-carbon sugar (pentose) plus an organic base derived from a purine or pyrimidine ring. Symbolically, this may be represented as

$$Base + Sugar \xrightarrow{H_2O} Nucleoside$$

$$Nucleoside + H_2PO_4^- \xrightarrow{-OH^-} Nucleotide$$

$$n - Nucleotide \xrightarrow{-(a-1)H_2O} Nucletic\ Acid$$

The number n is very large because the molecular weight of nucleic acid molecules usually runs in the millions. There are two types of sugar molecules included in nucleic acids. Throughout any one nucleic acid molecule, all the sugar residues are the same.

One sugar found in nucleic acids is called D-ribose, in which case the polymer is ribose nucleic acid (RNA); the other sugar possible is D-2-deoxyribose, in which case the polymer is deoxyribose nucleic acid (DNA).

The ring with carbon atoms at four of the corners is supposed, in this illustration, to appear to lie in one plane with the hydrogen and hydroxyl groups on bonds at right angles to the plane. Although this type of structure is often drawn, it is believed that the ring is not restricted to one plane but rather pleated, up and down, about the plane.

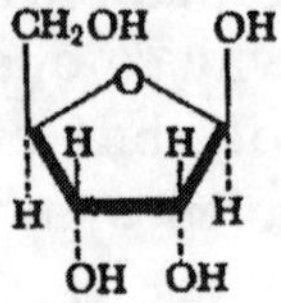

Fig. D-ribose

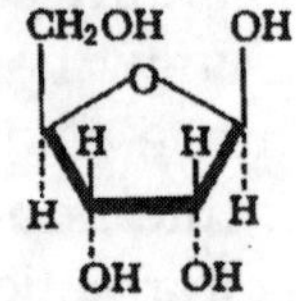

Fig. D-2-deoxyribose

The organic bases referred to above are derived from purine and pyrimidine rings. These rings have structures which can be shown as

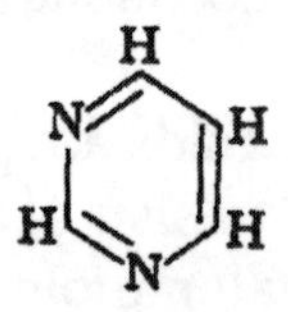

Fig. Pyrimidine

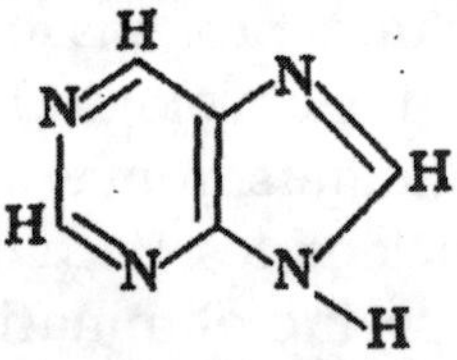

Fig. Purine

where the corners without letters in the ring are to be interpreted as carbon atoms). The two pyrimidines found in DNA are cytosine (C) and thymine (T), represented structurally by

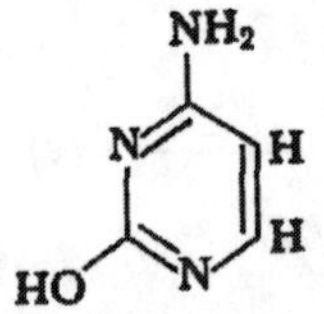

Fig. Cytosine

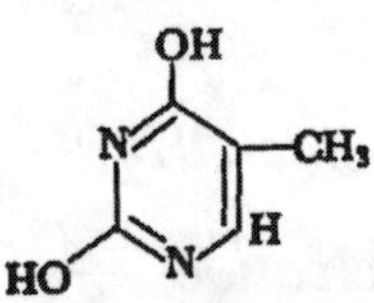

Fig. Thymine

In RNA, two pyrimidines are also found. These are cytosine and uracil. The latter has the structure

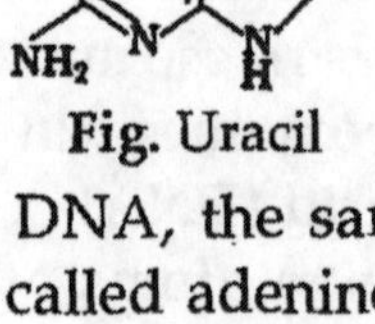

Fig. Uracil

In both RNA and DNA, the same two purine-derived bases occur. These are called adenine (A) and guanine (G). The structual formulas of adenine and guanine are

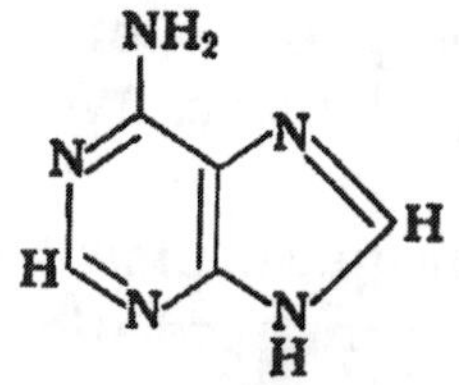

Fig. Adenine

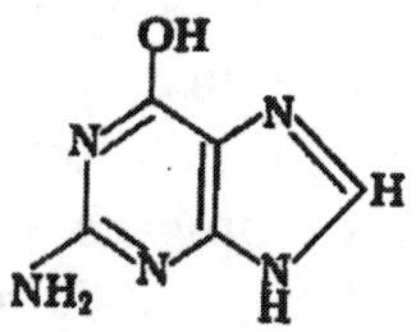

Fig. Guanine

Most genetic information in plant and animal cells is believed to be coded in DNA. This has only four monomers; these are nucleotides containing respectively A, T, C, and G. If it seems surprising that this alphabet is sufficient, it should be remembered that any English sentence can be written in Morse code which has three basic letters, a dot, a dash, and a pause.

X-RAY DIFFRACTION

The physical behaviour of molecules found in biological structures can be investigated from different points of view. One of the most fruitful of these has been an analysis of the atomic architecture as determined by X-ray diffraction patterns. The term "X ray" is used to refer to a beam of photons (electromagnetic radiation) formed by bombarding a metal target with electrons. These X rays are shorter in wavelength than other electromagnetic radiation referred to as visible and ultraviolet light.

The method of X-ray diffraction is a relatively new one in physical chemistry. X rays were discovered by Roentgen just before the start of this century. Quite a bit of simple X-ray crystallography was done between 1912 and 1920. However, accurate measurements of X-ray wavelengths and the corresponding studies of crystal structure have been possible only since about 1920. These studies profoundly affected scientists' ideas of the physical world at many different levels. The form of the periodic table, the exact values of the electronic charge e and of Avogadro's number N, and the arrangement of atoms in crystals and electrons within atoms, all have been based on X-ray measurements.

Although the diffraction of X rays by simple crystals, as NaCl, had been well studied for many years, the present interpretations of X-ray diffraction patterns of biologically interesting molecules were formulated since World War II. These were made possible by the same factor which is basic to so much of biophysics, namely the development of suitable electronic techniques. The detailed interpretation of X-ray diffraction data from complex molecules is possible only with the use of electronic analog computers and of high speed, digital electronic computers.

These studies of the diffraction of X-ray beams by biologically interesting molecules have influenced current ideas of the structure and action of almost all forms of biological compounds. The arrangement of the atoms within small molecules such as amino acids, purines, and sugars have been (or are being) determined. The chemical structural formula of certain antihistamines and the various isomers of vitamin A can best be investigated by their X-ray diffraction patterns. The helical structures of crystalline and fibrous proteins and of the genetic material, DNA, have been established from their diffraction of X rays.

The resolving power of an X-ray diffraction apparatus is much greater than that of a light microscope. In the light microscope, the limit of resolution is set by the wavelength of incident light employed. With X-ray diffraction patterns, no such restrictions exist. Using monochromatic X rays such as the Cu-K $_{á2}$ radiation, the wavelength A is 1.54 Å but interatomic distances can readily be measured with an error of less than 0.01 Å. This can be compared with a theoretical limit of resolution of 2×10^3 Å for blue light.

One may ask why a crystal has to be used rather than a single molecule, if the resolving power is indeed of the order of 0.01 Å whereas the covalent bond lengths average about 1.5 Å. Perhaps the most obvious answer is that it is impossible to hold one molecule in place. In addition, some of the X-ray photons will break molecular bonds. Because many molecules are present, breaking a few bonds does not have an appreciable effect on the average diffraction pattern. Perhaps the most important advantage of a crystal is that it restricts the scattered rays to a finite number of maxima, giving sharp, intense reflections.

One of the difficulties of X-ray diffraction studies is that one ends up with a photograph or graph with a number of spots of varying intensity. The problem of reconstructing the crystal and the spatial arrangements of the molecules from these spots has a simple solution only for crystals of very simple molecules, such as NaC1 or H_2O. For more complex molecules, a series of trial-and-error solutions is necessary. The analysis which follows is presented in the hope that those readers unfamiliar with this technique will acquire some idea of the problems involved.

Bragg showed that in treating X-ray diffraction by single crystals, one may regard the atoms as making up reflecting planes. A beam of X rays is shown incident on a single pair of such planes (although there will in general be many planes for any given crystal). There will be a maximum in the diffraction pattern if, and only if

$$\nu\lambda = 2d \sin \theta$$

where n is an integer. With monochromatic X rays, one set of planes, at most, will give a maximum for a given θ, and for any arbitrary θ there will probably be no maximum in the diffraction pattern. By rotating the X-ray beam around the crystal or by rocking the crystal about an axis perpendicular to the plane of the paper. The latter alternative is more practical and is often employed. Another type of diffraction pattern, called a Laue pattern, avoids this problem by using heterochromatic X rays of many wavelengths incident in a fixed direction. However, because λ will not be known, one cannot find the distance between reflecting planes from a Laue pattern.

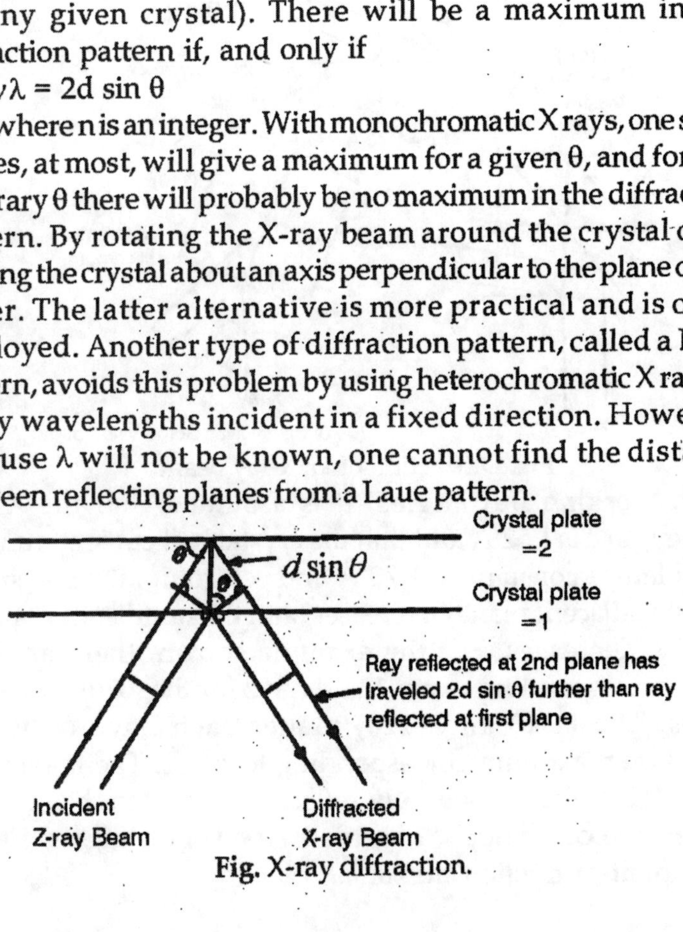

Fig. X-ray diffraction.

The dotted lines show perpendiculars to the wavefront. For re-enforcement, the rays reflected at planes 1 and 2 must travel distance which differ by a whole number of wavelengths. This will be fulfilled if

$$\nu\lambda = 2d \sin \theta$$

where n is an integer.

Several planes in a cubic crystal, each with a number of atoms per plane. These planes are numbered by Miller indices (hkl). The planes are spaced at varying distances. By and large, as the Miller

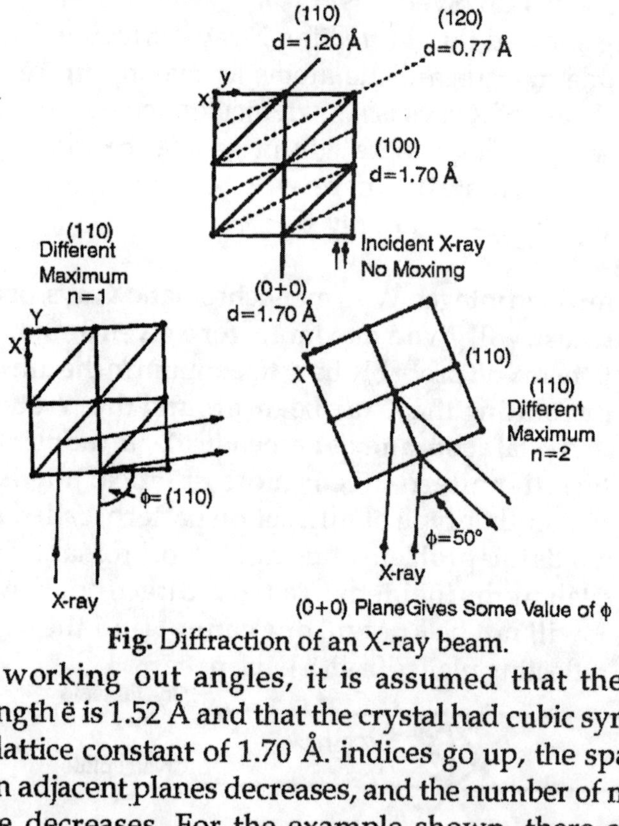

Fig. Diffraction of an X-ray beam.

In working out angles, it is assumed that the X-ray wavelength ë is 1.52 Å and that the crystal had cubic symmetry with a lattice constant of 1.70 Å. indices go up, the spacing d between adjacent planes decreases, and the number of maxima likewise decreases. For the example shown, there are two angles corresponding to n = 1 and n = 2 for the (010) and (100) planes. The (110) and (120) planes each have only one diffraction maximum corresponding to n = 1. The maximum for the (120) plane essentially reflects the incident beam back on itself and could not be observed. None of the higher planes will exhibit diffraction maxima.

However, planes not perpendicular to the xy plane, such as (101) and (011), will give maxima whose diffracted beams will not lie in the plane of the paper. To show all the planes with monochromatic X rays, a number of different schemes have been developed which lead to an easier interpretation of the Miller indices of the planes giving rise to a given reflection. For complex molecules, such as proteins and nucleic acids, it is necessary to use one of these schemes.

Instead of rocking the crystal, a powder of small crystallites can be used. Then the two-dimensional picture consists of a series of concentric circles. The powder pattern is harder to interpret but easier to obtain. It has been widely used in metallurgy and mineral sciences but has had few biological applications.

A sensitive test of an assumed atomic structure is to compare the relative intensities of the X-ray diffraction maxima observed with those computed from the model. The relative intensities are found by adding together the contributions of each atom, taking into account the phase differences due to the difference in pathlength to each atom. This is usually expressed by a crystal structure factor, F_{hkl}, for the beam perpendicular to the (hkl) planes. It can be found from

$$F_{hkl} = \sum_{n=1}^{N} \int_n e^2 \pi i \left(hu_n + ku_n + lw_n \right)$$

where

N	=	number of atoms in a unit cell of the crystal
n	=	a particular atom
fn	=	atomic structure factor defined below
(un, vn, wn)	=	coordinates of the nth atom expressed as fractions of the
		unit crystal lattice lengths

The atomic Structure Factor f is Defined by

$$f = \frac{\textit{Amplitude of the wave scattered by an atom}}{\textit{Amplitude of the wave scattered by an electron}}$$

In general, f depends both on the particular element (for example, Zn) and on the angle between the incident and the scattered beam; tables of f for various elements are available.

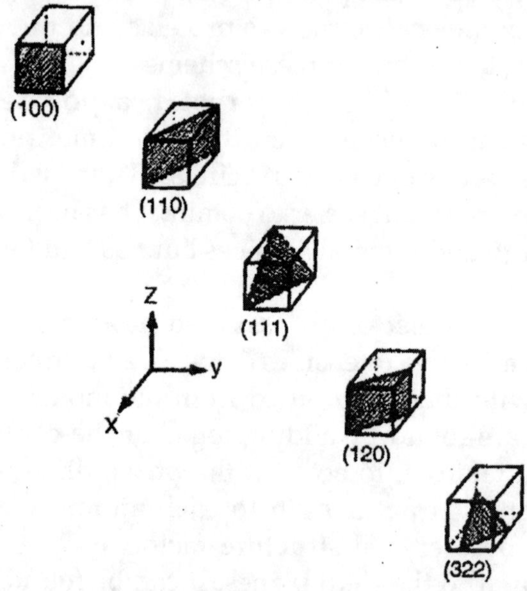

Fig. Miller Indices (hkl) for Some Crystal Planes Illustrated for Cubic Crystals.

The relative intensity will be proportional to |2|2. In a typical experiment, these are measured and the types of atoms present (and hence, the values of f $_{n)}$ are known. Thus, the assumed values of un, vn, wn can be used to compute |Fhkl|. It remains to adjust un, vn, wn for each atom until the final structure agrees both with chemical data and also with the X-ray diffraction pattern.

Complex crystals or even simple crystals of complex molecules give rise to complex diffraction patterns. The number of points necessary increases both as the number of atoms per molecule and also as the size of the unit cell of the crystal increase. In order to obtain useful information from these complicated diffraction patterns, it is necessary to know the relative intensities of the various maxima as well as their direction.

In interpreting diffraction by complex molecules, it is more convenient to deal with electron densities than with atomic positions. After the electron density has been mapped, the atoms may be located at the centre of the density maxima. From the preceding paragraphs, it may be seen that the crystal structure factor F_{hkl} can be defined by an absolute value $|F_{hkl}|$, and a phase angle a_{hkl} as

$$|F_{hkl}| = \frac{\text{Amplitude of the wave scattered by all atoms in the unit cell}}{\text{Amplitude of the wave scattered by an electron}}$$

a_{hkl} = phase difference between the wave scattered by the unit cell and the wave scattered by an electron at the origin

Adding the contribution of each electron as before

$F_{hkl} = \int\int\int$

where

ρ = electron density at u, v, w

Fourier series will recognize that Equation has the form of the coefficients of a Fourier series. Accordingly, one may rewrite it as

$$\rho(u,v,w) = \sum_h \sum_k \sum_l |F_{hkl}| \cos\{2\pi(hu + ku + lw) + a_{hkl}\}$$

Thus, if one can guess the values of a_{hkl} and can measure a sufficient number of intensities $|F_{hkl}|2$, then one can map the electron density ñ and hence, locate all the atoms. This is called a Fourier synthesis.

The problem of correctly guessing the phases either in Equation has intrigued mathematically minded crystallographers. The general procedure is to guess phase values and then keep adjusting these to give sharper and sharper electron-density contours. If one gets on the right track, these contours define atoms of the types known to be present and at reasonable distances from other atoms to which they can be bonded.

(The bonds can often be found by the methods of classical organic chemistry.) In the final analysis, phase guessing is very similar to working a crossword puzzle or solving a murder mystery story, and as in these, one finds, if successful, an answer which is no longer a guess.

Various schemes have been developed for the initial-phase guessing. One of the most successful involves placing a heavy atom such as I or Br within the molecule. The heavy atom diffracts more strongly than the others, so it may be located first. To do this, Equation is used, setting f_i to zero for all but the heavy atom. Once it is located, approximate values for many of the phases can be determined at once. With these as a starting point, one adjusts these phases and the others to give more and more sharply defined electron-density contours. The final solution is an accurate determination just as in the crossword puzzle.

For certain crystals, the unit cell is symmetric about the centre and, therefore, it can be shown that all the a hkl 's have the value 0 or ð. These are only two choices, but if 100 points are used there are 2^{100} or about 10^{30} possible sets of guesses. By the use of heavy-atom substitution, this hopelessly large number may be reduced to a mere few billion. Protein and nucleic acid crystals are not even symmetric about the centre of the unit cell, so that the problem is more difficult when using these molecules.

The entire adjustment of phase values and recomputing the F's and ñ is a lengthy, tedious process. With an electrical desk calculator and a protein crystal, this would take many human lifetimes. With electronic computers, it has been possible to find the details of the structural arrangements of the atoms within many smaller biological molecules.

PROTEIN STRUCTURE

One class of proteins, the globulins, could exist in either a fiberlike or a globular state. Most proteins do not have these two alternatives but, rather, are only fibrous or only globular. X-ray diffraction studies have been applied to both types of protein structure with varying degrees of success.

The problem of determining the structure of fibrous proteins is quite different from that of crystalline proteins. If a fiber were made up of little crystalline regions all lined up, then one should obtain spots similar to those from a single crystal. If these were slightly disoriented, the spots would become arcs. If the crystallites were completely randomly disordered, the spots would become circles similar to those of a powder pattern. Early investigators took many pictures of X-ray diffraction patterns of fibrous materials but no one understood the results.

Around 1930, Astbury studied many protein and nucleic acid fibers. He showed that the proteins gave rise to two types of X-ray diffraction patterns called a and b. He recognized the a-configuration as a folded or more dense structure, and the ß pattern as due to a stretched structure. The double-bonded oxygen is slightly negative, whereas the hydrogen on the nitrogen is somewhat positive. Alternate rows of polypeptide chains were postulated to be held in place by the attraction between these two, called hydrogen bonding.

Astbury showed that wool and hair changed from an a to a ß form on stretching. He postulated that in muscular contraction, the proteins changed from a b to an a form and this was generally accepted until about 1945. His models represented tremendous strides in understanding the structure of proteins but left many questions unanswered. His attempts to fit DNA and RNA patterns to his models were unsuccessful. Many spots in the diffraction pattern and the absence of other spots could not be explained in terms of his models. Thus, they represented a good first guess.

The current interpretations of X-ray diffraction patterns of fibers date back to about 1947. At that time, angles and configurations of many of the amino acids were investigated. Pauling and Corey, in 1951, showed that if they drew a polypeptide chain with the known bond angles on a piece of paper and twisted it into a helix, the various turns could hydrogen bond to one another. Astbury and others had tried helical models but always with an even number of amino acids per turn;

Pauling realised that this was a mistake. Pauling and Corey demonstrated that a helix with 3.7 amino acid residues per turn, a diameter of 6.8 Å and turns spaced 5.4 Å apart would be permitted by the observed bond angles. Left-handed a-helices are also possible. However, the diffraction pattern for X rays due to such a helical structure was more complex than that for any of the simpler models. Cochran, Crick, and Vand carried out a theoretical study of the predicted X-ray pattern for helical structures. They showed that it was necessary to orient the X-ray beam at an oblique angle to the fibers. Whereas the simpler molecules and crystals involved such terms as cos 2p (hx + ky + lz + a), the helical structures involved Bessel functions of the distance from the centre of the helix.

The a-helical structure of Pauling and Corey fits very well to the diffraction patterns observed for synthetic polypeptides. It is generally accepted that many natural fibrous proteins occur as helices because their X-ray diffraction patterns are of the a-type which is similar. However, many unexplained spots are present.

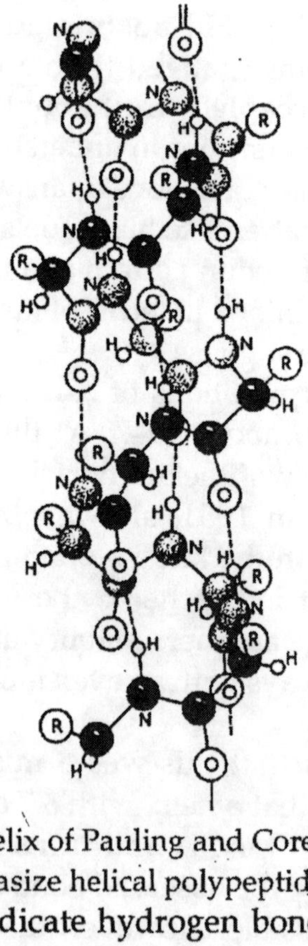

Fig. The a-helix of Pauling and Corey Redrawn to Emphasize helical polypeptide chain.

Dotted lines indicate hydrogen bonding between = O of

one turn with the H—N of the turn below. This is a closepacked structure. The data for the fibrous protein collagen can be fitted by a model involving a fiber made of three a-helices twisted around each other in a helical fashion. By far, the most direct demon. stration of the a-helix is in the globular protein myoglobin, discussed subsequently. In addition to their a-helix, Pauling and Corey made pleated sheet models of proteins similar to the β model of Astbury but did not restrict the peptide bonds to one plane. Both this and the a-helix have retained the, ideas of the a structure being compressed and the β stretched out, and also of hydrogen bonds being responsible for holding the shape of the protein fibers.

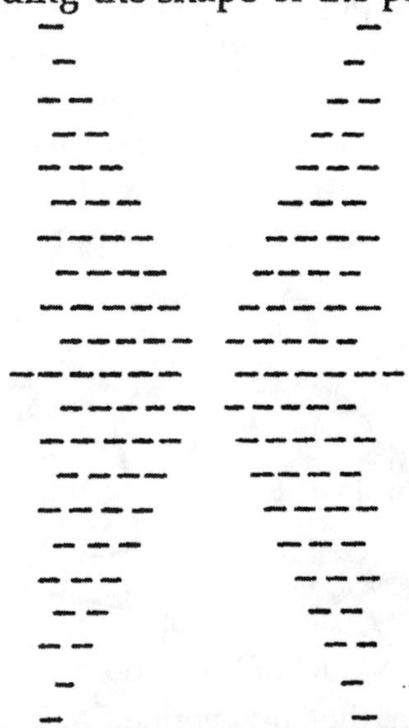

Fig. Diffraction Pattern of a Helix. Notice the Clear area in the Centre of the Pattern.

They are superior to Astbury's earlier models in fitting known bond angles and in their agreement with the experimental results of X-ray diffraction studies.Perhaps the most studied crystal has been that of the blood protein, hemoglobin. However, the structure of the similar but simpler protein, myoglobin, was worked out to a resolution of about 6 Å before much progress was made with hemoglobin. Myoglobin is a red pigment similar to hemoglobin but occurring in muscle rather than, blood.

It is believed to function by buffering the oxygen concentration within the muscle. Myoglobin has a molecular weight of about 16,000, very low for a typical protein. This corresponds to 153 amino acid residues, that is, about 1,200 atom's other than hydrogen in each myoglobin molecule. It means that to locate all of these atoms in the molecule, one would need to measure the intensity and to guess the phases of perhaps 20,000 diffraction spots. The results of analyzing and adjusting the phase for about 400 diffraction spots for myoglobin crystals, substituted with heavy atoms, showed there were two myoglobin molecules per unit cell of the crystal and located the polypeptide chains and the iron-containing heme group within the myoglobin molecule.

This procedure was then repeated to include 9,600 diffraction spots which showed the electron density of the myoglobin molecule with a resolution of about 2 Å. This is not quite sufficient to indicate the separate atoms.

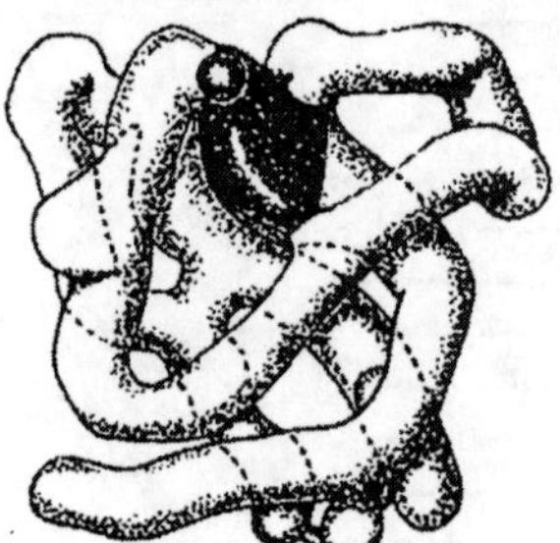

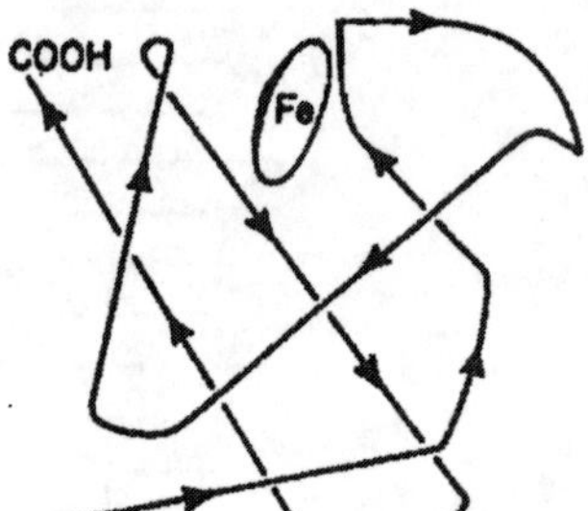

Fig. Kendrew's Model of Myoglobin. (a) General shape of the Polypeptide Chain.

This resolution is sufficient to confirm that the major part of the myoglobin molecule consists of righthanded a-helices. To fit the single polypeptide chain of myoglobin into one globular molecule, it must be bent and twisted at various corners. Where this occurs, the a-helical form is lost usually for 3 or 4 amino acid residues. There is also one group of about 13-18 amino acid residues not in the form of an a-helix.

A photograph of Kendrew's model of myoglobin, built to represent the structure which would give the 400 diffraction spots used. In addition, electron spin resonance measurements were used to locate the iron atoms in the heme group.

However, the latter data were misinterpreted so that the heme group was tilted at the wrong angle. For comparison, shows the form of the polypeptide chain revealed by the 9,600 diffraction spot study. All of the chains are shown by the latter study to be hollow, cylindrical tubes of the form expected for helices.

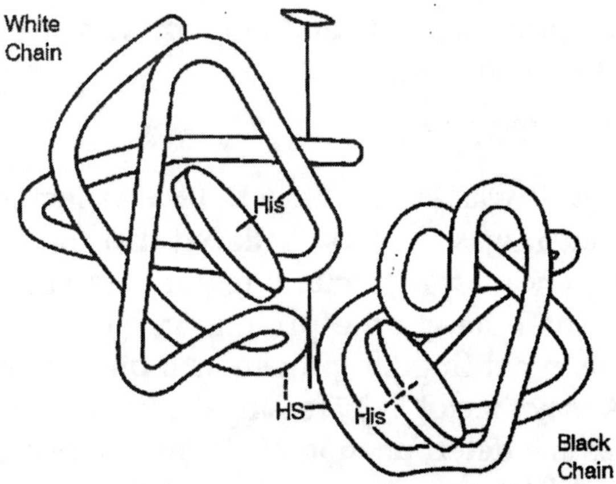

Fig. Perutz's Model of Hemoglobin.

The straight chain portions are right-handed á-helices. It was necessary to use four different substitutions of heavy atoms to check the results and obtain suitable starting points for phase guessing.

Similar studies of hemoglobin (molecular weight about 65,000) have shown that it consists of four subunits, each with a heme group. These studies at a resolution of 5.5 Æ showed that each subunit of hemoglobin is a continuous polypeptide chain folded around itself in a form very similar to that of the myoglobin molecule. There are two identical pairs of subunits in each molecule. These are shown in black and white in the model, which summarizes the X-ray diffraction studies of Perutz and his co-workers.

The myoglobin and hemoglobin studies involved three new ideas not used in the 1920's in X-ray determination of inorganic crystalline structure. The first was the technique of the substitution of heavy atoms into the unit cell (or isomorphous replacement, as it is called). The second involved the use of high-speed computers to adjust the phases until the electron density postulated and the diffraction spots observed were consistent with one another. The third novel technique was the use of electron spin resonance to locate the iron atoms. Everything indicates that with the development of higher-speed computers, with better programming for electronic computers, and with the preparation of increasing numbers of crystalline proteins, more and more protein structures will be studied by these methods.

NUCLEIC ACID STRUCTURE

Although it was known that DNA was included in the chromosomes, it was formerly believed that the nucleic acid was only significant for structural reasons. The current views completely interchange the protein and DNA role in the chromosome, regarding the protein as a protective agent for the DNA. A major factor in increasing the significance assigned to DNA was the determination of a satisfactory steric model by Crick and Watson. Their interpretation made use of the theory referred to in the last section for X-ray diffraction by helical structures.

Crick and Watson showed from the X-ray diffraction patterns that DNA consists of two antiparallel helices. The spiral is very large, having a diameter of 18 Æ and a spacing between turns of 34 Æ. Thus, it is a wide chain with lots of room for other molecules to fit in, if they are of the proper shape. On the basis of chemical data and size considerations, Crick and Watson showed that the chains were made up of -sugar'!phosphate-sugar'!phosphate- and so forth, units.

Between the two chains, as rungs along a helical step ladder, were strung pairs of hydrogen-bonded bases of the form purine—H—pyrimidine. These rungs are about 11 Æ long. This type of unit is repeated into a long double chain. The entire double chain is then twisted to form the double helix, with about 10 rungs per turn.

The pairs of bases fit across the chain as rungs being supported in the middle by hydrogen bonds. It is necessary that each pair of bases fit very exactly. Measurements based on X-ray diffraction patterns of crystals of the purine and pyrimidine bases have shown that these do indeed fit, provided that adenine (A) is paired with thymine (T), and guanine (G) with cytosine (C). If this is the case, one should have the relative concentrations of organic bases in DNA related as [A]/[T] = [G]/[C] = 1.0 This relationship had been verified for all DNA and was one of the pieces of evidence used by Crick and Watson to construct their model.

Top, general picture of the double helix, with the phosphate-sugar combinations making up the outside spirals and the base pairs the cross-bars; middle, a somewhat more detailed representations: phosphate (P), sugar (S), adenine (A), thymine (T), guanine (G), cytosine (C), and hydrogen (H); bottom, detailed structure showing how the space is filled with atoms: carbon (C), oxygen (O), hydrogen (H), phosphorus (P) and the base pairs. Gamow tried to explain protein formation in terms of the location of residues along the DNA chain. In the outlined cross, G, T, and C are independent, but the A is determined by the T. Crosses of the nature outlined could determine amino acid order. However, several difficulties were present. A model of this nature made in three dimensions with actual bond angles, and so on, failed to show any good place to fit the amino acids.

The model used each purine or pyridine in two crosses so that there would be an appreciable influence of one amino acid on the next; this has not been found in natural proteins. Finally, other biological data favour control of the synthesis of RNA by DNA. In turn, RNA is involved in protein synthesis. In spite of its obvious limitations, Gamow's model indicated information could be coded in DNA.

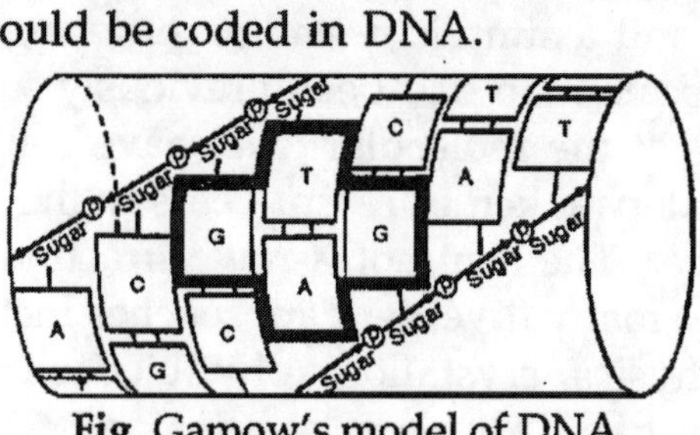

Fig. Gamow's model of DNA.

X-ray diffraction patterns have been studied for RNA as well as DNA. The structure of RNA is more complicated than that of DNA. The ratio of purine to pyriraidine is not one in

RNA. The RNA chains may be branched. Its X-ray diffraction pattern suggests a single-chain helical symmetry. The detailed interpretation of this pattern still has not been accomplished.

In spite of this failure to determine the details of the structure of the RNA molecule, X-ray diffraction and electron scattering have been used to locate the nucleic acid within the RNA-type virus particles. This may be done for any of a considerable number of viruses which have been crystallized. The very existence of crystals implies detailed ordering at the atomic level, and accordingly a definite X-ray diffraction pattern. A model of tobacco mosaic virus (TMV) constructed to fit X-ray diffraction data. More recent models of TMV reveal that the RNA is held in a large vacuous helix similar to the DNA helix but with quite different physical dimensions. Instead of being supported by hydrogen bridges, the RNA helix in TMV is held in place and stabilized by the surrounding protein molecules which actually intermesh with the RNA.

Another plant virus whose gross structure has been determined is bushy stunt tomato virus (BSV). The particles of BSV are spherical, but, just as TMV, consist of a nucleic acid core plus a surrounding envelope of protein. The X-ray study of the structure of living matter is a fascinating and growing field. It has made possible discoveries of the steric form of complex high polymers such as proteins and nucleic acids, both of which are responsible for many of the properties of all living systems. These studies, based on X-ray diffraction, have revealed both a complexity that was previously beyond imagination, and a simplicity and an ordering of atoms on a much larger scale than had been previously suspected.

Analyses at the molecular level have contributed many of the major steps taken in recent years to the understanding of living matter. The result of X-ray diffraction studies alone; rather, a great many divergent approaches including those of chemistry, physics, crystallography, biochemistry, genetics, and virology have all been synthesized to elucidate the structure of proteins and nucleic acids. The task is far from complete and one may anticipate continued development of these techniques.

Chapter 11

Molecular Action of Ionizing Radiations

The effects of ionizing radiation on living matter can be considered at many different levels. The most complex is that of the entire animal or plant. Exposure of entire organisms to ionizing radiation may result, depending on the dose, in induction of cancers and mutations, radiation sickness, and even death. Although health physicists must learn what these complex responses are and at what dosages they are likely to occur, the complexity prohibits complete understanding at this time. Eventually, it is believed that the responses of higher animals and plants to ionizing radiation will be explained in terms of effects occurring at the cellular level.

Particular stress was given to changes induced in the chromosomes. The cellular effects of ionizing radiations can be explained in part in terms of molecular changes. The effects on the molecular level, and in particular on the natural high polymers. Further logical extensions of this process of abstraction would be to consider the interactions of ionizing radiations with small molecules, atoms, and electrons. Biologically important effects of ionizing radiations in living matter often involve two molecular species, proteins and nucleic acids.

This is particularly true of the chromosomal changes, for chromosomes are composed of proteins plus nucleic acids. Both types of compounds are complex high polymers, occurring naturally. The effects of ionizing radiation on simpler high polymers are easier to interpret; these are discussed first.

This information is then applied to the changes induced in proteins and nucleic acids by ionizing radiation.

Several different types of radiation produce ionization in all high polymers, both natural and synthetic. These same types of radiation produce ionizing effects in biological cells. Some of the physical characteristics of such radiations, which include α, β and ã rays, proton, neutron, and deuteron beams, and X-ray photons. The action of all of these radiations is associated with ionization and the breaking of bonds. In certain cases, such as dried protein films, the local ionization in the polymer is the only active process. However, when proteins and nucleic acids are in solution, the ionization of the solvent plays an important role.

POLYMERS, PROTEINS, AND DNA

Chemically, both proteins-and nucleic acids belong to a general class of compounds called polymers (or high polymers). A polymer is made up of a repeating type unit (the monomer), which is duplicated again and again; the repeating units making up proteins are called amino acids. Similarly, there are repeating units making up nucleic acids; their monomers are called nucleotides. The amino acids are joined together by peptide bonds to form proteins. Because a molecule of water is eliminated for each peptide bond formed, it is customary to state that the amino acids condense to form the proteins.

Nucleic acids, too, are condensation-type polymers, a molecule of water being eliminated for each nucleotide joined to the chain. Both proteins and nucleic acids are complicated high polymers in that the long chains consist of a mixture of different types of residues or monomers arranged in a definite but complex pattern. It is easier to interpret the effects of ionizing radiations on simpler synthetic polymers, made up of one or at most two types of monomers, than on the more complex proteins and nucleic acids. A study of the radiation damage to these simpler polymers provides orientation toward the types of effects to be expected when proteins and nucleic acids are exposed to ionizing radiations.

Studies have been made of the effects of ionizing radiations on many different types of synthetic high polymers. Two used as examples in the next section are polyethylene and polyisobutylene.

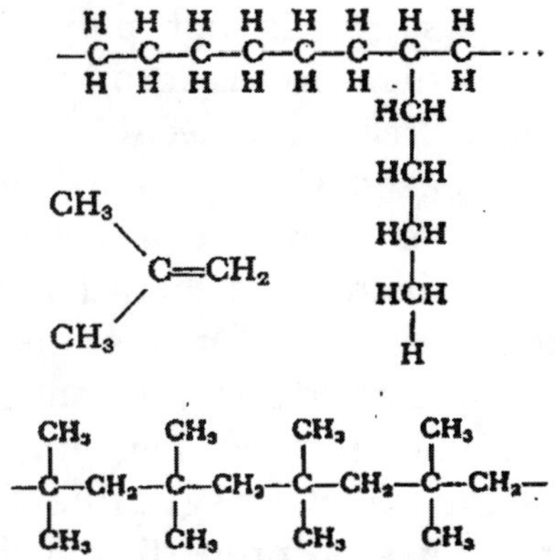

Fig. Structure of Polyethylene and Polyisobutylene.

When the monomers (ethylene) molecules are added together to form polyethylene, it is possible to have carbon atoms surrounded by only seven electrons. Those atoms with unpaired electrons are called free radicals.

Many free radicals are extremely reactive; they are believed to be responsible for continuing a chain-type reaction once polymerization is started. Free radicals are also responsible for some of the effects of ionizing radiations on high polymers. In fact, the primary action of the ionization is to knock an electron or proton away from the polymer leaving the latter as a free radical.

RADIATION DAMAGE TO SYNTHETIC HIGH POLYMERS

When high polymers are irradiated in solution, two different types of damage may occur. The first is that the polymer molecule itself may be irreversibly altered owing to the direct action of the radiation in producing ionizations within the polymer molecule.

The second type of damage possible is due to indirect effects; these result from the reactions of the polymer with the free radicals formed in the solvent by the ionizing radiation. In dilute solutions, it is very likely that the indirect effects may be the more important ones. On the other hand, in the solid state, the direct actions of the radiation on the polymers are the only important type.

As a result of both direct and indirect damage, two major changes are found in synthetic high polymers. The first is called crosslinking, which means forming bonds between chains. It results in increased molecular weight, increased elastic moduli, increased transparency, and decreased solubility. The other type of effect, called scission, consists of breaking bonds along chains. It is characterized by the exact opposite of the effects described for crosslinking.

With both crosslinking and scission, a small molecule such as H_2 or NH_3 is often eliminated. This third effect of small molecule elimination causes negligible changes in the molecular weight or physical properties of the polymers as compared to the changes due to scission and crosslinking. However, such small changes can completely alter the physiological actions of a protein. Probably all high polymers undergo both scission and crosslinking when irradiated. However, in some, such as polyethylene the crosslinking is the predominant effect.

Its physical characteristics are changed in a fashion that indicates only the crosslinking. By contrast, polyisobutylene, shows only the effects associated with scission when it is exposed to ionizing radiation. The effects of ionizing radiations on all polymers are usually enhanced if oxygen is present. For example, if polyethylene is bombarded in the absence of oxygen, crosslinking occurs.

If oxygen is present during the bombardment, considerably more crosslinking takes place. Likewise, the scission in polyisobutylene is greater if it is exposed to radiation damage in the presence of oxygen rather than in the absence of oxygen. The enhancement of the effects of bombardment is manyfold larger than expected from the ionization of the oxygen molecule by the radiation. The increased effects must result from the reactions of oxygen with the free radicals and other less stable forms produced by the ionizations.

In a molecule such as polymethylene, which is the same as polyethylene except that it has no side chains, there is no reason to suppose one carbon bond to be weaker than another. In other molecules, such as polyisobutylene, the carbon atoms attached to four other carbon atoms seem to be the weak links in the chain. In more complex polymers, there is usually one weakest bond. It appears that the extra energy imparted by the ionizing radiation is often carried to this spot. Thus, iso-octane breaks preferentially at one particular bond.

Another example of the transport of the ionization energy can be found in solutions. Some molecules which fluoresce because of the formation of free radicals will do so when the solution is irradiated, even if the ionization occurs in the solvent at a distance of 50 A or more away. Similarly, most aromatic-ring compounds tend to stabilize free radicals and can protect polymers. This occurs either if the aromatic compounds are placed in solution with the polymer or if they are incorporated into the polymer, as in polystyrene. The extra energy imparted by ionization may be transferred within molecules or between molecules. In some cases, this involves charge transfer, but in others it does not. Exactly how the latter happens is not always understood.

There are situations in which the extra energy imparted by the ionizing radiation is not transferred. For instance, colour changes and electrical-resistance changes in polyethylene, both of which are due to the presence of free radicals, may remain for weeks or even months after irradiation, provided oxygen is excluded. In this case, there is no doubt that the extra energy is not transferred through the polymer. No general rule exists to predict why the energy is transferred in some cases and not in others.

In the more complex natural polymers,the phenomena which should be considered in natural polymers are: direct and indirect action; crosslinking and scission; the elimination of small molecules; the role of oxygen; energy transfer; and damage protection. Dried protein films are simpler than protein solutions or biological cells, in that only direct effects are possible. From the variation of molecular destruction in dried films with dose rate, it is possible to compute a sensitive volume for the molecular species being studied.

INACTIVATION OF DRIED PROTEIN FILMS

When dried protein films are subjected to ionizing radiation, the molecules are irreversibly altered. Because no solvent is present, the changes observed are of necessity direct ones in the protein itself. Protein films are usually tested for molecular changes after redissolving them in suitable media (usually buffered water). Tests of the physical and chemical properties of proteins are far more sensitive to any small change whatsoever than are those used on synthetic polymers. However, in most cases, even though a change can be detected, it is not possible to determine whether crosslinking or scission has occurred within the protein molecule.

The polypeptide chains making up the protein appear to be so crosslinked that either scission or additional crosslinking can occur without altering the molecular weight. In other words, protein changes may be detected with a high sensitivity, but the tests yield no knowledge whatsoever of the intramolecular alterations. An exception to this inability to distinguish scission from crosslinking is the molecule, hemocyanin, an iron-containing, oxygen-transport pigment found in snails and other invertebrates. When hemocyanin, molecular weight 7,000,000, is exposed to certain chemical agents such as urea, it undergoes scission, being split reversibly into eight pieces of equal molecular weight. Ionizing radiations also split hemocyanin; however, they split the molecule irreversibly in half.

One very sensitive criterion for physical changes in a protein is its solubility. Another is its isoelectric point (that is, the pH of the medium at which the protein molecule will not migrate in an electrical field). If the protein is an enzyme, the enzymatic activity is a very sensitive indication of its physical and chemical condition. Finally, other proteins react with specific antibodies; in this case, the protein is called an antigen. Its antigenicity may be altered after irradiation. Studies on a large number of dried protein films have shown all of the preceding changes. No dried protein films are completely unaltered by ionizing radiation.

The elimination of small molecules observed with synthetic high polymers can be readily demonstrated for proteins. When either the monomers (amino acids) or their high polymers (proteins) are exposed to ionizing radiation, a number of small molecules are eliminated. These include NH_3, CO_2, and CO. In the case of single amino acid molecules, the elimination of a smaller molecule represents a scission or a breaking of bonds. An amino acid which does not eliminate NH_3, CO_2, or CO during irradiation is cysteine, which contains a sulfhydryl group, –S–H.

In the presence of O_2 or other oxidizing agents, two cysteines may be oxidized and then may unite to form one cystine. In proteins, there are some indications that such a high fraction of the free –S–H groups are oxidized that energy appears to be transferred preferentially to these sulfhydryl groups from other parts of the molecule. (In pure cysteine, in the absence of any oxidizing agent, H_2O_2 and H_2S are formed under the action of ionizing radiations. Proteins apparently stabilize the free radicals in the cysteine residues so that little or no H_2S is released.)

In spite of the indeterminancy of the nature of the changes in most dried proteins, it is possible to investigate energy transfer directly. If one assumes target theory and determines a damage or inactivation versus dosage curve. If energy transfer may occur throughout the protein molecule, this critical volume should be the entire molecule. It is likewise conceivable that energy transfer could occur in only part of the molecule or, at the other extreme, energy transfer may take place between adjacent molecules. These conditions could lead to critical volumes which would be small or large, respectively, as compared to one protein molecule.

All three possibilities—critical volume equal to, smaller than, and larger than one molecule—have been observed by using dried films of different proteins. The case of the critical volume equal to the molecular volume has been observed for DNA-ase, invertase, and many other proteins, as well as for one smaller molecule, penicillin, molecular weight 600. Although energy transfer throughout the molecule is a sufficient condition for the equality of the molecular and critical volumes, it is not a necessary condition. It is possible that any of a variety of small changes in various parts of the molecule would all lead to the same conclusion of protein damage. Thus, finding a critical volume which equals a molecular volume suggests that energy transfer may take place throughout the entire molecule but does not prove it.

In contrast, a critical volume smaller than the molecular volume proves that energy transfer cannot occur throughout the entire molecule. This is the case for catalase which has a critical volume, as determined by tests using enzymatic activity, corresponding to one-half its molecular weight. (If urea is added to the test medium, it is found that the critical volume corresponds to the entire molecular weight.) Bovine serum albumin is a more extreme example. When a monolayer is irradiated, the critical volume determined by antigenic tests corresponds to onetenth its molecular weight. This implies that severe damage to ninetenths of the molecule can leave the antigenic activity unaltered. Thus, both catalase and bovine serum albumin may be partially damaged without inactivating the remainder of the molecule.

Finally, some protein films are able to transfer excitation energy from one molecule to the next. Insulin films show a critical volume corresponding to a molecular weight of 23,000, whereas the usual value for the molecular weight is 12,000. All linked together sufficiently tightly for energy transfer to occur between them. The digestive enzyme, trypsin, is similar in that it has a molecular weight of about 20,000, but its critical volume corresponds to a molecular weight of 34,000. Thus, energy apparently may be transferred from one molecule to the next. If enzyme-substrate complexes or enzyme-inhibitor complexes are exposed as dried films to ionizing radiations, the enzymes are inactivated. In these cases, it is found that one ionization anywhere within the complex is sufficient to inactivate the enzyme. Here, energy transfer occurs across the bonds holding the complex together.

Experiments similar to those with protein films can be done with dried viruses. These have all shown that the critical volume for genetic changes is comparable to the volume indicated by the nucleic acid content. Moreover, they have shown the complexity of the virus, that any gross criterion such as virus multiplication cannot be conveniently described in terms of Equation. Rather, to use Equation, one has the alternatives of measuring the rate of any genetic change whatsoever, or else comparing the rate of production of the same mutation as indicated by the methods presented.

It should be noted that the inactivations of dried protein and virus films by bombardment with ionizing radiations have indicated several features of the physical properties of proteins and viruses. The similarities between proteins and synthetic high polymers are emphasized by the similarity in the types of radiation damage, whereas the greater complexity of the proteins is emphasized by the greater sensitivity of tests for changes in the proteins. Differences between proteins are demonstrated by the various ranges of energy transfer; in some, the excess energy may be transferred throughout the entire molecule, in others it is restricted to part of the molecule, and in a third group energy transfer could occur between loosely bound pairs of molecules. Finally, the accepted role of the nucleic acids in virus multiplication is in accord with the results obtained through the irradiation of virus particles.

INDIRECT EFFECTS ON PROTEINS AND NUCLEIC ACIDS

When solutions of proteins and nucleic acids are exposed to ionizing radiations, the changes observed are very similar to those found for dried protein films. However, if the concentrations are low enough, there can be little doubt that many of the effects are indirect, arising from ionizations occurring in the solvent. This is always true for proteins at physiological concentrations. In the case of nucleic acid solutions, and particularly of DNA solutions, the question of direct or indirect effects is not so clear cut.

A great deal of evidence supports the direct target theory for DNA in chromosomes and bacteriophages. It shows that ionizations, leading to changes within a critical volume corresponding to about four nucleotides, are sufficient to alter genetic character. It is possible to believe therefore that energy is not transferred very far along the DNA polymer chain. However, the primary effect in living cells could be due to the formation of free radicals in the cellular water in a small volume close to the DNA molecule.

Because water plays an important role in damage to proteins in solution and may also be significant for DNA changes, effects of ionizing radiations on water have been carefully investigated. The primary process may be represented by

$$H_2O \rightarrow HO\cdot + H\cdot$$

where the dots signify free radicals. Alternative forms are also possible, for example,

$$H_2O \rightarrow \begin{Bmatrix} H_2O^+ \\ + e^- \end{Bmatrix} \begin{array}{l} \nearrow HO^+ + H\cdot \\ \searrow H^+ + HO\cdot \end{array}$$

$$e^- + H_2O \rightarrow H_2O^- \begin{array}{l} \nearrow OH^- + H\cdot \\ \searrow OH\cdot + H{:}^- \end{array}$$

These may then recombine in various forms to produce reactive molecules such as H_2O_2 and H_3O. All these reactive forms are known to alter proteins and DNA. The probability of producing H_2O_2 in water is greatly increased in the presence of dissolved oxygen.

Many experiments have confirmed that damage to both proteins and DNA in solution is much smaller either in the absence of oxygen or in the presence of a protective agent, which is more likely to react with the H_2O_2, and free radicals than are the proteins and nucleic acids. Protective agents of this type can also reduce damage due to direct effects of the ionizing radiations because the extra energy may be transferred to the protective agent from the protein or nucleic acid.

Among the most effective compounds of this nature are a group containing both $-S-H$ or $-S-S-$ groups and amino groups $-NH_2$, such as cysteamine and cystamine. These compounds protect not only proteins and nucleic acids, but also synthetic condensation-type polymers such as polymethacrylic acid. Other protective agents with very different structures have also been used, such as ß-aminoethylisothiuroniumbromide-hydrobromide.

In terms of critical volumes, the net effect of the protective agent will be to reduce the critical volume far below the molecular volume. Various theories have been developed to explain the action of the protective agents on proteins as due to stabilizing terminal —S—H and —S—S— groups. However, because the same agents also protect synthetic polymers which do not contain sulfur, it appears unjustified to focus too much attention on one particular type of bond.

Studies have been conducted on the relative sensitivities of many amino acids, proteins, and nucleic acids to ionizing radiations. These have led to quite lengthy tables, but no one has succeeded in relating this information to protein structure. The relative sensitivities of different proteins can be indicated by a G value; this is the number of molecules altered per 100 ev (of a fixed type of irradiation). Some typical values are shown in the accompanying table which indicates the widespread range of G values.

The values in the table show that as the complexity of the molecule increases, there is some tendency for the molecule to become less sensitive to ionizing radiations. Thus, adenine is more sensitive by itself than when combined with ribose to form the nucleoside, adenosine. It in turn is more sensitive than the nucleotide, adenosine monophosphate (adenylic acid). Complete nucleic acids are another order of magnitude less sensitive. However, the variation of sensitivity with size is not always observed. Some proteins are almost as sensitive as a typical small molecule, whereas others are far less sensitive, so that no general rule can be maintained.

Exposures of proteins and nucleic acids to ionizing radiations have shown that all of these molecules may be altered with sufficient dosages. The similarity between the damages observed in proteins and nucleic acids, and those in synthetic high polymers, has been emphasized. These damages include: direct and indirect effects; crosslinking and scission; elimination of small molecules; and energy transfer. All play an important role in the irradiation of living matter by ionizing particles.

Protein enzymes	G
Yeast alcohol dehydrogenase	3.4
Carboxy peptidase	0.55
Hexokinase	0.033
Catalase	0.009
Nucleic Acids	
DNA	0.0039
RNA	0.0072
Nucleotides	
Adenylic acid (AMP)	0.161
Nucleosides	
Adenosine	0.196
Purine	
Adenine	0.676
Other Small Molecules	
DPNH	1.7
Ethyl alcohol	5.9
Coenzyme A	9.2
Glutathione	10.7

The data on the irradiation of dried protein films have also contributed to our knowledge of the physical properties of these proteins. However, the net return of knowledge has been very small, being limited to an indication of energy transfer and the relative sensitivity of the proteins to ionizing radiations. In this respect, the methods of X-ray analysis. A quite different aspect of proteins, namely enzymatic activity, have shown far more about the physiological role of proteins than has been discovered in experiments using radiation damage.

Chapter 12

Molecular Nanotechnology

Some pragmatic readers may be wondering what is so special about molecular nanotechnology, that existing or anticipated biotechnology could not accomplish just as well? After all, biotechnology is already an established medical capability. It has real applications and real products already on the market. Reflecting upon the future possibility of sophisticated mechanical medical nanorobots equipped with powerful nanocomputers, in 1989, one well-known cryobiologist mused that "what is not clear is just what need we would have for such devices." The simplest answer, is that each of the three contemporary branches of "nanotechnology" offers something of unique value to the practice of medicine.

Nanoscale materials technology has already found widespread use in medicine, including biocompatible materials and analytical techniques, surgical and dental practice, nerve cell research using intracellular electrodes, biostructures research and biomolecular research using near-field optical microscopy, scanning-probe microscopy and optical tweezers, and vaccine design, and also many 20th century bulk chemical and biochemical manufacturing techniques along with much of classical pharmacology.

As for "biotechnology," the original meaning of this word contemplated "the application of biological systems and organisms to technical and industrial processes". In recent times, the field has expanded to include genetic engineering and now takes as its ultimate goal no less than the engineering of all biological systems, even completely artificial organic living systems, using biological instrumentalities.

The third branch, molecular nanotechnology, takes as its purview the engineering of all complex mechanical systems constructed from the molecular level – potentially offering new tools for medical practice. Observes G.M. Fahy, "the difference between nanotechnologists and biotechnologists is that the former do not restrict themselves to the biological limitations of the latter, and they are much more ambitious about the kinds of accomplishments that they want to achieve."

Doctors can utilize solutions to medical problems from all three approaches. As noted earlier, 80-90 Percent of medical complaints resolve themselves via natural homeostatic processes, or without the necessary involvement of active biotechnological or molecular-nanotechnological agents. But by employing biotechnology, the range and efficacy of treatment options greatly increases.

With molecular nanotechnology, the range, efficacy, comfort and speed of possible medical treatments again expands enormously. Molecular nanotechnology is essential when the damage to the human body is extremely subtle, highly selective, or time-critical, or when the damage is very massive, overwhelming the body's natural defences and repair mechanisms.

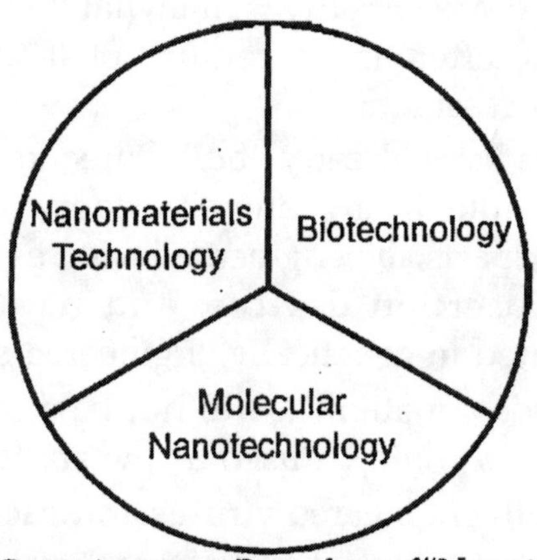

Fig. Three Contemporary Branches of"Nanotechnology"

At every difficulty level, most classes of medical problems may be resolved with varying efficacy within the homeostatic/ nanomaterials, biotechnological, or molecular-nanotechnological approaches. As the chosen technology becomes more precise, active, and controllable, the range of options broadens and the quality of the options improves. Thus the question is not whether molecular nanotechnology is required to accomplish a given medical objective. In many cases, it is not though of course there are some things that only biotechnology and molecular nanotechnology can do, and some other things that only molecular nanotechnology can do. Rather, the important question is which approach offers a superior solution to a given medical problem, using any reasonable metric of treatment efficacy. For virtually every class of medical challenge, a mature molecular nanotechnology offers a wider and more effective range of treatment options than any other approach.

It is quite possible to imagine an advanced biotechnology that uses an engineered white cell, fibroblast, or macrophage chassis, energized by native oxygen and glucose and modified mitochondrial powerplants, driven by pseudopodia, cilia or flagella, communicating and navigating via biochemical signals, and even incorporating onboard digital biocomputers to make microscopic biorobots. Principal arguments favouring the biotechnology approach for medical purposes are:

- That we are already somewhat familiar with such systems, after half a century of intensive molecular biology research;
- That we have already "built" precursor systems, such as a whole living amoeba constructed from five distinct parts, bioengineered viruses and bacteria as DNA insertion devices, and natural replication stimulated in genetically engineered starter microbes;
- That biocompatibility will not be a major issue since fibroblasts could be used as the starter material;
- That both engineered viruses and bacteria are already in wide commercial and research use; and
- The greater complexity of self-repair in mechanical systems, should it be needed. Many believe that the development pathway to early biorobots may be considerably shorter than for the mechanical

nanorobots of the molecular nanotechnology approach, for which in 1998 not a single working prototype yet existed, even in research laboratories.

It is also possible to imagine a molecular nanotechnology that uses mechanical nanorobotic systems. Such systems will have many constitutional differences from biological-based systems. Mechanical systems will transport parts, materials, energy and instructions via fixed channels, whereas most biological systems operate by diffusion.

Mechanical systems will have structures constrained by specific geometries, whereas biological systems have structures defined by patterns of containment and interconnection — the shape of a membrane compartment in a cell matters less than its continuity and the contents of the volume it defines. Mechanical systems will be deterministically manufactured by operations analogous to manual construction, whereas engineered ribosomes will self-assemble via diffusion and stochastic matching of complementary parts; in other words, biology uses recipes, while mechanical systems use blueprints. Cells grow, with their parts adapting to one another; mechanical nanorobots may be constructed from parts of fixed structure.

Biology uses self-repair; mechanical systems generally do not — largely because self-repair will not be needed in molecular mechanical systems, whose designs may be made more simple by relying upon high component redundancy. These and other differences imply a number of important advantages that mechanical-based medical systems will enjoy over biological-based medical systems, which, taken together, strongly suggest that predominantly mechanical nanosystems may be the approach of choice for a mature medical nanotechnology. The advantages of molecular nanotechnology are many:

SPEED OF MEDICAL TREATMENT

Doctors may be surprised by the incredible quickness of nanorobotic action when compared to the speeds available from fibroblasts or leukocytes. Normal homeostatic processes such as dermal wound repair via natural fibroblasts may require weeks to run to completion. Typical fibroblast movements occur at 0.1-1 microns/sec but mechanical nanomanipulators can operate at 1-10 cm/sec speeds or faster, a speed advantage of 4-5 orders of magnitude. Even the strongest biological fibres have a failure strength 3 orders of magnitude below the strongest mechanical fibres.

Biological cilia beat at ~30 Hz while mechanical nanocilia may cycle up to ~20 MHz, though practical power restrictions and other considerations may limit them to the ~10 KHz range for most of the time. Flexible mechanical surfaces can complete a morphing motion in ~0.1 millisec, compared to the ~100 millisec snapback time for pinched red cell membrane, again a thousandfold advantage in speed.

Thus we expect that mechanical therapeutic systems can reach their targets up to ~10,000 times faster, all else equal, and treatments which require ~10^5 sec for a biological system may need only ~10^2 sec for a mechanical system; tachyiatria improves both patient and physician comfort. In either biological or mechanical systems, large numbers of devices of comparable physical size may be employed to do the work, so numbers alone cannot offset the mechanical speed advantage.

POWER DENSITY AND TRANSDUCTION

Biological cells typically employ power densities of 10^3-10^4 W/m^3, with maximum densities of ~10^6 W/m^3 in honeybee flight muscle cells and bacterial flagellar motors By contrast, nanomechanical power systems can produce power densities of 10^9-10^{12} W/m^3, an advantage of 10^3-10^8 for mechanical over biological systems. By 1998, conducting polymer-based actuators generated 20-100 times the force for a given cross-sectional area as mammalian skeletal muscle.

Additionally, amoebic locomotion in motile cells requires diffusion-limited cytoskeletal disassembly and reassembly to achieve movement; mechanical motility systems may employ simple cable-pulling, winches, or ratchets, which are faster and more direct. Some biological energy transducers are reversible, but muscle contraction is irreversible – not only cannot muscle actively re-expand, but stretching it doesn't make it produce much useful chemical energy. In contrast, electric motors can be run backwards to generate electricity, forcing pistons makes them pump, and loudspeakers can be used as microphones.

SUPERIOR BUILDING MATERIALS

Typical biological materials have tensile failure strengths in the 10^6-10^7 N/m^2 range, with the strongest biological materials such as wet compact bone having a failure strength of ~10^8 N/m^2, all of which compare poorly to ~10^9 N/m^2 for good steel, ~10^{10} N/m^2 for sapphire, and ~10^{11} N/m^2 for diamond and carbon fullerenes, again showing a 10^3-10^5 advantage for mechanical systems that use nonbiological materials. Nonbiological materials can be much stiffer, permitting the application of higher forces with greater precision of movement, and they also tend to remain stable over a larger range of temperature, pressure, salinity and pH. Proteins are heat sensitive in part because much of the functionality of their structure is due to noncovalent bonds involved in folding, which are broken more easily at higher temperatures; in diamond, sapphire, and many other rigid materials, structural shape is covalently fixed, hence is far more temperature-stable.

Most proteins tend to become dysfunctional at cryogenic temperatures, unlike diamond-based mechanical structures. Biomaterials are not ruled out for all nanomechanical systems, but represent only a small subset of the materials that can be used in nanorobots. Mechanical systems can employ a wider variety of atoms and molecular structures in their design and construction, with novel functional forms that might be difficult to implement in a biological system such as steam engines or nuclear power.

As another example, an application requiring the most effective bulk thermal conduction possible should use diamond, the best conductor available, not some biomaterial with inferior thermal performance.

NONDEGRADATION OF TREATMENT AGENTS

Diagnostic and therapeutic agents constructed of biomaterials generally are biodegradable in vivo, although there is a major branch of pharmacology devoted to designing drugs that are moderately non-biodegradable – anti-sense DNA analogues with unusual backbone linkages and peptide nucleic acids are difficult to break down. However, suitably designed nanorobotic agents constructed of nonbiological materials are not biodegradable.

An engineered fibroblast may not stimulate an immune response when transplanted into a foreign host, but its biomolecules are subject to chemical attack in vivo by free radicals, acids, and enzymes. Even "mirror" biomolecules or "Doppelganger proteins" comprised exclusively of unnatural D-amino acids have a lifetime of only ~5 days inside the human body. Nonbiological materials such as diamond and sapphire are highly resistant to chemical breakdown or leukocytic degradation in vivo, and pathogenic biological entities cannot easily evolve useful attack strategies against these materials.

CONTROL OF NANOMEDICAL TREATMENT

Present-day biotechnological entities are not programmable and cannot be switched on and off conditionally during task execution. A digital biocomputer, while possible in theory, represents a considerable conceptual departure from the usual biological paradigm.

Even assuming that a digital biocomputer could be installed in a fibroblast, and that appropriate effector mechanisms could be attached, such a system would necessarily have slower clock cycles, less capacious memory per unit volume, and longer data access times, implying less diversity of action, poorer control, and less complex executable programmes than would be available in nanoscale electromechanical computer systems.

The mechanical approach emphasizes precise control of action, including control of physical placement, timing, strength, structure, and interactions with other entities. The biological approach emphasizes the use of poorly controlled natural structures, needlessly sacrificing huge blocks of the available functionality and design space.

NANODEVICE VERSATILITY

Mechanical systems can readily incorporate biological elements if necessary, but artificial biological systems can incorporate nonbiological materials such as carbon nanotubes or diamond/sapphire structural elements only with difficulty, in part because biology has a more limited repertoire of "effector" mechanisms.

Artificial biological systems cannot easily incorporate nonbiological materials where desired because natural biological assembly methods make no provisions for these materials either in the coded instructions in DNA or in the attachment chemistries. Rebuilding or reconstructing the human body with nonbiological components, or augmentation of human body function with unnatural abilities or features, will be very difficult or impossible to achieve using purely biotechnological means.

AVOIDING OVERSPECIALIZATION

M. Krummenacker notes that one of the most glaring shortcomings of bacteria and other naturally occurring molecular machinery – when viewed as systems subject to further engineering – is the rather limited range of molecular substrates they can utilize. Bacterial enzymes are highly specialized devices, with very narrow substrate specificities. Thousands of different enzymes are needed in each organism, and the substance classes capable of digestion are limited to some sugars, various amino acids including proteins that can be degraded by excreted proteases, lipids, and a few other smallish oxygen-functionalized carbon molecules such as glycerol and ethanol.

Some bacteria can metabolize CO_2 and a handful of aromatic compounds, but there is a vast range of organic chemicals that most bacteria cannot degrade or manufacture. A much smaller set of substantially more general molecular tools can probably be designed using the mechanical approach;

mechanosynthesis can fabricate and assemble, or disassemble, a far wider range of molecular structures than are available to the cellular machinery of life.

FASTER AND MORE PRECISE DIAGNOSIS

The analytic function of medical diagnosis requires rapid communication between the injected devices and the attending physician. If limited to chemical messaging, biotechnology devices will require minutes or hours to complete each diagnostic loop. Nanomachines, with their more diverse set of input-output mechanisms, can outmessage the results of in vivo reconnaissance or testing literally in seconds. Such nanomachines can also run more tests of greater variety in less time. Mechanical nanoinstruments, including molecule-by-molecule disassemblers, will make comprehensive cell mapping and cell interaction analysis possible. Bacterial resistance can be assayed at the molecular level, allowing new treatment agents to more easily be composed, manufactured and immediately deployed.

MORE SENSITIVE RESPONSE THRESHOLD FOR HIGH-SPEED ACTION

Unlike natural systems, an entire population of nanobiotic devices can be triggered globally by just a single local detection of the target antigen or pathogen. The natural immune system takes $>10^5$ sec to become fully engaged after exposure to a systemic pathogen or other antigen-presenting intruder. A biotechnologically enhanced immune system that can employ the fastest natural unit replication time will require $\sim10^4$ sec for full deployment post-exposure. By contrast, a nanobiotic immune system can probably be fully engaged in at most two blood circulation times, or $\sim10^2$ sec.

MORE RELIABLE OPERATION

Engineered macrophages would probably individually operate less reliably than mechanical nanorobots. Many *pathogens, such as Listeria monocytogenes* and *Trypanosoma* cruzi, are known to be able to escape from phagocytic vacuoles into the cytoplasm; while biotech drugs or cell manufactured proteins could be developed to prevent this, nanorobotic trapping mechanisms can be more secure. Proteins assembled by natural ribosomes typically incorporate one error per $\sim 10^4$ amino acids placed; current gene and protein synthesizing machines utilizing biotechnological processes have similar error rates. A molecular nanotechnology approach will improve error rates by at least a millionfold. Mechanical systems can also incorporate sensors to determine if and when a particular task needs to be done, or when a task has been completed.

Finally, it is unlikely that natural organisms will be able to infiltrate mechanical nanorobots or to co-opt their functions. By contrast, a biological-based robot could be diverted or defeated by microbes that can piggyback on its metabolism, interfere with its normal workings, or even incorporate the device wholesale into their own structures, causing the engineered biomachine to perform some new or different function than was originally intended.

There are many examples of such co-option among natural biological systems, including the protozoan mixotrichs found in the termite gut that have assimilated bacteria into their bodies for use as motive engines, and the nudibranch mollusks that steal nematocysts away from coelenterates such as jellyfish and incorporate the stingers as defensive armaments in their own skins, a process which S. Vogel has called "stealing loaded guns from the army."

VERIFICATION OF PROGRESS AND TREATMENT

Using a variety of communication modalities, nanorobots can report back to the attending physician, with digital precision, a summary of diagnostically- or therapeutically-relevant data describing exactly what was found, and what was done, and what problems were encountered, in every cell visited. A biological-based approach relying upon chemical messaging is necessarily slow with limited signaling capacity. Also unlike mechanical nanorobots, biotechnological systems generally cannot monitor their own functions while working, and, except for a few highly specialized DNA proofreading systems, cannot directly inspect their work while it is in progress or after it is finished.

MINIMUM SIDE EFFECTS

Almost all drugs have side effects, such as conventional cancer chemotherapy which causes hair loss and vomiting, although computer-designed drugs have high specificity and relatively few side effects. Carefully tailored cancer vaccines under development in the late 1990s were expected unavoidably to affect some healthy cells. Even well-targeted drugs are distributed to unintended tissues and organs in low concentrations, although some bacteria can target certain organs fairly reliably without being able to distinguish individual cells. By contrast, mechanical nanorobots may be targeted with virtually 100 Percent accuracy to specific organs, tissues, or even individual cellular addresses within the human body.

Such nanorobots should have few if any side effects, and will remain safe even in large dosages because their actions can be digitally self-regulated using rigorous control protocols that affirmatively prohibit device activation unless all necessary preconditions have been, and continuously remain, satisfied. G.M. Fahy has noted that these possibilities could transform "drugs" into "programmable machines with a range of sensory, decision-making, and effector capabilities [that] might avoid side effects and allergic reactions...attaining almost complete specificity of action....Designed smart pharmaceuticals might activate themselves only when, where, and if needed."

Additionally, nanorobots may be programmed to excuse themselves from the site of action, or even from the body, after a treatment is completed; by contrast, spent biorobotic elements containing ingested foreign materials may have more limited post-treatment mobility, thus lingering at the worksite causing inflammation when naturally degraded or removed.

REDUCED REPLICATOR DANGER

Drexler points out that living systems are evolved

systems, while nanomechanical replicators would be designed: "The former are shaped to serve the goal of their own survival and replication in a natural environment, whereas the latter will be shaped to serve human goals, perhaps in an artificial environment." Genetic engineering involves not design of replicators from scratch, but tinkering with the molecular machinery of existing bioreplicators. Since bioreplicators were not designed, they are not necessarily structured in ways that lend themselves to complete understanding, and processes based on diffusion and matching allow complex nonlocal interactions that can be hard to trace.

Bioreplicators can be crippled, but having evolved in nature, they resemble systems that can survive in nature. Typically, they are able to exchange genetic information with wild organisms, raising the possibility of the introduction of new, unconstrained replicators in the natural environment. Having evolved to evolve, they have a capacity for further evolution – to serve their own survival, not human goals. Even when stripped of key pieces of DNA to interfere with its replication powers, a live attenuated AIDS vaccine can slowly recover its virulence and can attack immune cells. R. Bradbury suggests that artificial biorobots could incorporate multiple fail-safe mechanisms including required external essential nutrients or suicide suppressors, self-destruct triggers, countdown timers like telomeres, and engineering to reject foreign DNA, but it remains logically easier for a system with inchoate capacity to evolve to resume doing so, than for a system which has never had this capability to spontaneously develop it.

In contrast, nanomechanical replicators will be designed from scratch and thus will differ fundamentally from biological systems. The parts and structures of designed mechanical systems will be known, and the relationships among their parts will also be designed and fixed. More important, nanoreplicators will be fundamentally alien to the biosphere, unrelated to anything that has evolved to survive in nature. Certainly the capacity to fail can appear by accident, and emergent capabilities cannot be completely ruled out, but engineering experience shows that the ability to perform complex organized activities does not normally appear spontaneously. Also, and purely as a geometrical consideration, adding a new part inside a densely organized geometric structure typically requires changes in the relative positions of many other parts, and hence corresponding adjustments in design. Adding a part inside a densely organized topological structure typically leaves topologies unchanged — room can be made by stretching and shifting other parts, with no change in their essential design, hence permitting easier modification to biological structures by exogenous agencies.

Note that mechanical medical nanodevices need not be capable of replication. There is no requirement for replication in vivo; such replication would be needlessly dangerous, and adding this capability would reduce effectiveness in carrying out the primary medical task. Analogously, viral vectors employed in genetic therapies are modified to be "incapable" of replication.

ASSURED PATENTABILITY

Microscopic biorobots, unavoidably derived from natural biological material, may someday be deemed unpatentable under a general prohibition on "genetic colonialism" or other emerging legal doctrines. In contrast, mechanical nanorobots, being fully-artificial and designed machines, should always be patentable provided they satisfy the customary legal criteria